THE GREAT SENEY FIRE

A HISTORY OF THE WALSH DITCH FIRE OF 1976

By Gregory M. Lusk

ISBN: 978-0-578-95908-5
Copyright Registration Number: TXu 2-372-282
Library of Congress Control Number: 2023911739

Snowsnake Press
900 Third Street
Hancock, MI 49930

Book Layout by Gregory M. Lusk
Cover by Gregory M. Lusk
Fire Progression Map by Cory Luoto

The Great Seney Fire/ Gregory M. Lusk. —1st ed.
Printed and bound in the USA

First Paperback Printing-March 2023
First Hardcover Printing-January 2024

Snowsnake Press

DEDICATION

To the combined efforts of the more than one thousand Department of Natural Resources and Federal firefighters, cooperators, and support staff that responded to the challenges of suppressing the largest fire in Michigan since 1908. The remarkable effort stands out as an example of what dedicated professional people can do together as a team. They all worked long and hard with unity and distinction to finally extinguish the Great Seney Fire.

"A spark neglected has often raised a conflagration."

- Quintus Curtius Rufus -

Roman historian, 1st century CE

CONTENTS

LIST OF FIGURES

INTRODUCTION

By the middle of summer in 1976, it was evident that fire danger was mounting significantly in the Upper Peninsula, and a record-setting drought was becoming protracted. A lightning strike in late July and a unique set of events set in motion the most destructive and costly fire in Michigan since 1908. Aside from its sheer magnitude, the "Great Seney Fire" stands out as an example of cooperation and a remarkable team effort. Questions will be asked about the fire long into history, and we can argue about the answers, but no one can argue about the superb job that was done in suppressing the fire, a job done by highly skilled professional people dedicated to the conservation of Michigan's and the nation's natural resources. Many lessons were learned by the organizations involved; hopefully, this account will help others remember and understand this small but important piece of Michigan history.

CHAPTER ONE

NATURAL HISTORY

The eastern parts of the Upper Peninsula and the Lower Peninsula are underlaid by a series of sedimentary rock layers known as the Michigan Basin. These rock formations, consisting mainly of shales, limestones, and sandstones, were deposited on the bottom of ancient seas that covered Michigan on and off for millions of years. Kiril Spiroff, Professor of Geology, Michigan Tech University, described the geologic development of the upper peninsula as follows:[1]

> *Lofty peaks of active volcanoes once topped Michigan; shallow seas once covered it; deserts, jungles, and later ice ruled the landscape. The Upper Peninsula bedrock formations include some of the oldest rocks known on earth. Gravels, sands, marls, and silts were deposited in shallow seas or on great outwash plains that existed in this region a billion or more years ago. Life was not abundant in those ancient times, but humble forms of the plant and animal kingdoms were present. Some of these lowly organisms formed reefs that are like deposits of a much later date. After great thicknesses of rock formations were laid down, the district became the site of a profound crustal disturbance. The rock beds were contorted and broken, high mountains were formed, and great masses of molten rock surged within the earth's crust and broke out to the surface in the form of volcanoes. In the center of Lake Superior, where deep cold water now conceals the remnants of this activity, large fissures poured forth lava that flowed as far as the present shores of this great lake. Intermittently, layers of gravel and sand spread over*

1 Spiroff, Prof. Kiril, Geologic History of Michigan, July 1-2-3-4, 1960, Mich. College of Mining & Technology Midwest Federation of Mineralogical & Geological Societies Twentieth Annual Field Trip Convention, "*Souvenir Bulletin of the Jaspilite*"

lava flows. Spent by all this activity, the area sank, leaving the outer part of the volcanic series as rims around the present Lake Superior. A million years or more after the Keweenawan epoch's eruptions, the Michigan Basin began to develop. During Paleozoic times, this large basin was slowly filled with sediments from nearby landmasses. By the end of Paleozoic time, the basin became a broad floodplain of a low swampy nature. Life had made enormous strides in development. Fishes, crustaceans, and mollusks were abundant. Plantlife was very conspicuous. The luxuriant vegetation of tree ferns, giant mosses, and horsetails growing in the warm moist climate transformed this low land into jungle swamps exceeding those today.

While they undoubtedly roamed on Michigan territory, nobody has ever turned up a single dinosaur bone in Michigan. From the Permian period (299-252 million years ago) to the end of the Neogene (23-2.6 million years ago), erosion occurred, and no sediments were deposited

Source: Dr. Randall Schaetzl, Michigan State University

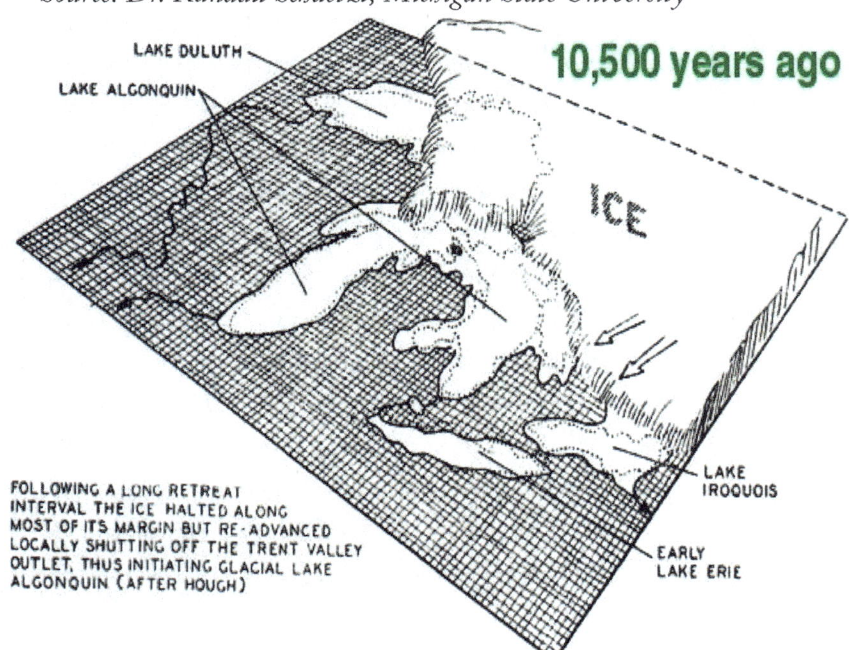

Figure 1. Glacial Lake Algonquin

locally, leaving behind no rocks or fossils. This period encompasses the entirety of dinosaurs' existence on the planet.[2]

As vegetation accumulated on the surface, it formed peat composed of partly decayed stems, leaves, and woody materials. After the disappearance of the last of the Paleozoic seas, Michigan may not have been submerged again. However, the State was covered several times by enormous sheets of ice. These glaciers modified the appearance of the land. As they moved ahead, they grooved and polished the resistant rock outcrops. As the glaciers retreated, their load was deposited chaotically, leaving hills, gravel beds, and shallow lakes.[3] The entire region around Seney and the central Upper Peninsula is an extensive, almost level sandplain produced by the deposition of glacial outwash from the last ice age.

The lands comprising the Seney National Wildlife Refuge (NWR) present an area of seemingly little geological variation in comparison with more scenic areas along the shores of Lake Superior and Lake Michigan. The broad flat lands of the Refuge reflect a subtle but highly complex geologic history. Little topographic relief exists on the Refuge; the elevation ranges from 794 feet above sea level in the north to 640 feet in the southeast, and grades S 15° E at 16-32 feet per mile.[4]

FORMATION OF THE SENEY SWAMP

Glacial Lake Algonquin (Figure 1) inundated most of the eastern and southern Upper Peninsula, parts of the northern Lower Peninsula, and the Tahquamenon and Manistique's marshy lowlands. The level plain from Seney to Shingleton, Schoolcraft County, was the bed of

---

2 Serba, John, mLive.com, *Meet Michigan's Most Bizarre Prehistoric Creatures*, https://www.mlive.com/entertainment/2017/08/michigans_wildest_prehistoric.html, Posted Aug 13, 2017, updated May 20, 2019

3 Serba, John, *Meet Michigan's Most Bizarre Prehistoric Creatures*,

4 Urbanek, Richard P. and Bookhout, Theodore A., *"Nesting of Greater Sandhill Cranes on Seney National Wildlife Refuge"*, North American Crane Workshop Proceedings. 325. http://digitalcommons.unl.edu/nacwgproc/325, 1992

Source: Dr. Randall Schaetzl, Michigan State University

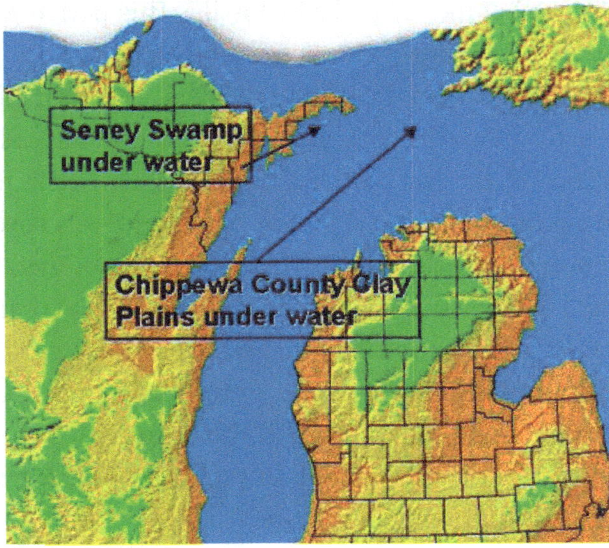

Figure 2. Seney swamp under Lake Algonquin

Lake Algonquin (Figure 2).[5] The swamps of the eastern UP are the remnants of that lake.[6] After these waters receded, long ridges of sandy gravel interrupted what would otherwise be great expanses of flat land. Many dunes occurred along their inland slopes, far from any body of water. These lands have areas of heavy clay interspersed with stretches of water-washed sand and gravel. They were once lake beds; the ridges of gravel were beaches; the flat stratified sand and gravel plains and clay plains with a veneer of sand or gravel were fillings of old bays of ancient long-departed lakes and the deltas of rivers flowing into them.[7] The small sand knolls that dot the flat plain are "extinct" dunes that likely formed about 9,500 years ago.

In the 1960s, researchers noticed, first in aerial photographs, a

5-7 Schaetzl, Dr. Randall, *Geography of Michigan & the Great Lakes*, GEO 333, Part II. Physical Geography, Spring Semester, 2005, Michigan State University, http://geo.msu.edu/extra/geomich/part two-B.html, and http//geo.msu.edu/extra/geomich/glacial-lake_algonq.html

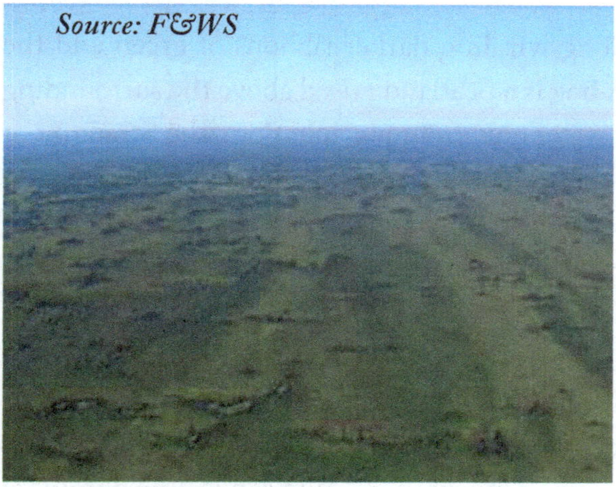

Figure 3. The Strangmoor bog

distinct patterning of part of this area, a "strangmoor" or a "string bog," which are usually called "patterned fens." A fen is a peatland fed by groundwater that has been in contact with mineral bedrock or soil. The pattern in the photos shows the conspicuous wetlands' bandings with many small circular dark areas of small pines on the old dunes (Figure 3). Between the thin linear areas

Figure 4. Strangmoor bog location

of bog ridges are the long winding darker ribbons of green and the lighter treeless bogs.[8] A bog is a peatland raised above the surrounding landscape that receives water only from precipitation. The Strangmoor Bog at Seney is one of the southernmost examples of a string bog. Its protection is essential while simultaneously creating challenging fire behavior when its deep fuels are in a drought (Figure 4).

SOILS

Within the Seney Sand Lake Plain, 100 to 200 feet of glacial drift generally cover the bedrock. The soils on the Seney NWR are generally level to somewhat sloping mucks, peats, and sands. The dominant mucks are interspersed with sand ridges and knolls in such an intricate pattern that the two soils have been mapped together as a complex of Carbondale muck and Rubicon sand (dune phase). The muck has accumulated on the wet sandy plain at a depth of 3 to 9 feet. The material is a dark brown, spongy, felt-like muck, which is more decomposed than peat soils and generally contains a higher percentage of mineral matter. The natural drainage is extremely poor in the mucks and excessive in the sands on the ridges and knolls. This complex covers most of the Refuge. Wet sands underlie the entire area.[9]

Along the Manistique River Valley, the Driggs River, and the other tributaries draining the Refuge, the soils are predominately sands and sandy loams. These soils are well or excessively drained and lie on slopes that are level to steeply sloping. The soil surface consists of forest litter, underlain by gray sandy loam or fine sandy loam, with coarser sand beneath the loam.[10]

---

8 The U.P.'s Natural Landmark – *Strangmoor Bog*-The Unique Formation of Seney's Natural National Landmark Conservation Gateway, Nature Conservancy, https://www.conservationgateway.org/ConservationByGeography/NorthAmerica/UnitedStates/michigan/projects/Documents/Landscape%20Stewardship%20Stories/14.StrangmoorBog.pdf posted 7/18/2017, accessed May 4, 2022

9-10 *Wildland Fire Management Plan*, Seney National Wildlife Refuge, Department of Interior, Fish and Wildlife Service, 2008

Most of the Seney swamp drainage water has a high iron content, as evidenced by iron concretions and deposits or mounds of bog iron on the surface, especially where the land has been burned over and by deposition of iron along drainage ditches. Most of this iron is probably derived from the iron oxide-bearing sands and sandstones, which underlie a considerable part of it and are transported as a bicarbonate solution. It is probably deposited as siderite in the presence of organic matter without access to air or as limonite in the presence of air. You can also see reddish-brown tints in the water from the iron and algae, which have concentrated in spots and settled to the bottom, creating layers of "bog iron."[11] While of no commercial importance, plates of bog iron can be found near the origin of the 1976 Walsh Ditch Fire, possibly acting as a ground rod, drawing lightning to it in the electrical storm blamed for its start.

PREHISTORIC PLANT AND ANIMAL LIFE

Fossil pollen records indicate that Michigan's vegetation's composition and distribution have been in slow and constant flux since the last glacial recession. Species migrated south and east of the ice sheet as changing climatic and soil conditions allowed. The first arrivals were pioneering arctic tundra-adapted plants that quickly colonized the barren landscape, south to north, between 17,000 and 13,000 years ago. This vegetation was succeeded by a spruce parkland that lasted until about 11,000 years ago when pine and most of the deciduous trees arrived in Michigan. Competition between species began, resulting in the expansion and contraction of different species' distribution according to their climatic tolerances. Climatic changes varied across

---

11 Foster, Z.C., *Soil Survey, Schoolcraft County, Michigan*, United States Bureau of Chemistry and Soils, United States Bureau of Plant Industry, Soils and Agricultural Engineering, United States Soil Conservation Service, United States Natural Resources Conservation Service, p. 38, (retrieved February 1, 2018) from Google Free E-books: https://books.google.com/books?id=QfjwAAAAMAAJ&pg=PA2&source=gbs_selected_pages&cad=3#v=onepage&q&f=false

the State, causing the species composition of forests in the Upper Peninsula to differ from those of the Lower Peninsula. Pine forests were dominant in most of the Upper Peninsula for much of the last 11,000 years in response to the prevailing warm, dry climate and the sandy soils. Starting about 6,600 years ago, the environment of the Upper Peninsula became cool and moist. Northern hardwood forests, dominated by birch, maple, hemlock, and beech trees, occupied the moist, fertile sites in the eastern half of the Upper Peninsula. In contrast, pine-dominated forests were present on the drier sites.[12]

Less common animal members of Michigan's fossil record include black bears, giant beavers, white-tailed deer, Scott's moose, muskrats, peccaries, and meadow voles. The most common mammals in Michigan's Pleistocene fossil record were elk, Jefferson mammoths, American mastodons, woodland muskoxen, and caribou.[13]

More commonly known as reindeer, the caribou existed in Michigan until roughly one hundred years ago. Lake Superior caribou are part of the species *Rangifer tarandus*, which includes reindeer in Eurasia, barren-ground caribou across the North American Arctic, and caribou in the boreal forests of Canada.[14]

Once the dominant deer species across the north, caribou roamed from Hudson's Bay to Mackinac Island in Michigan. They persisted for more than 1.5 million years in North America. When glaciers encroached, they found refuge in Appalachia's southern mountains; when temperatures warmed, they moved northward with the melting ice to fill the forests and islands along Lake Superior. They managed to thrive despite predators and hunting by Indigenous peoples. But

12 Hupy, Christina M., and Yansa, Catherine H., *The Last 17,000 Years of Vegetation History*, https://msu.edu/~yansa/2009a-Hupy%20&%20Yansa-CH%207-MI%20Book%20Chapter.pdf, 2009

13 *Paleontology in Michigan*, Wikipedia, https://en.wikipedia.org/wiki/Paleontology_in_Michigan November 13, 2021, accessed May 4, 2022

14 Myers, John; Northland Outdoors, *Last Lake Superior Caribou Surviving on Their New Island Homes*, http://www.northlandoutdoors.com/2020/04/06/last-lake-superior-caribou-surviving-on-their-new-island-homes/, April 6, 2020

they could not survive European settlement that brought over-hunting and development — farming, mining, and logging.[15]

The caribou's distribution and abundance along the southern edge of its range declined dramatically in the late 1800s and early 1900s. In the Lake Superior region, caribou were extirpated from the mainland of Michigan in 1912 and Isle Royale in 1928. They disappeared from Minnesota in the 1940s, but there were sporadic sightings of at least two caribou in extreme northeastern Minnesota during the winter of 1981-82. The current southern boundary of continuous woodland caribou distribution crosses Ontario at about 50 degrees latitude.[16]

Several caribou fossils dating to the Pleistocene era have been uncovered in Michigan. The species adapted to colder climates and followed the glacial retreat north to avoid extinction at the end of the Ice Age. Hunting, disease, and climate change are likely reasons for the caribou's reduced numbers.[17]

EARLY HUMAN ACTIVITY

The Paleoindians that arrived first were few and primarily hunters of big game. They generally made their camps at the juncture of flowing rivers. These hunters may have helped eliminate the large grazers, mammoths, mastodons, giant beavers, and similar herbivores that may have influenced early forest development. As the climate became warmer and drier, other human groups followed, about 7,500 years before present (BP). Between 7,000 and 6,000 BP, many more hunters were present. Some used copper spears, while others fashioned arrowheads from silicified sandstone. Upper Peninsula copper was traded extensively. A cooler climate prevailed after 3,500 BP, but warming

---

15 Myers, *Last Lake Superior Caribou Surviving on Their New Island Homes*

16 Gogan, Peter J. P., and Cochrane, Jean Fitts, *Restoration of Woodland Caribou to the Lake Superior Region*, U.S. National Park Service Publications and Papers. 11. http://digitalcommons.unl.edu/natlpark/11, 1994

17 Serba, John, mLive.com, *Meet Michigan's Most Bizarre Prehistoric Creatures*, https://www.mlive.com/entertainment/2017/08/michigans_wildest_prehistoric.html, Posted Aug 13, 2017, updated May 20, 2019

began around 3,000 BP with many fluctuations.[18]

Native Americans influenced the forests of the Upper Peninsula long before European settlement began. Their influence was most significant on fire-susceptible communities such as savanna, grassland, and pine forests. These were intentionally burned for ease of travel or to aid hunting. Sometimes considerable areas were cleared for cultivation or around settlements for fuelwood and to build stockades. In some locations, long "fences" of brush were constructed to funnel game to the hunters.[19]

The early European visitors appeared to have had little direct impact on Upper Peninsula forests. The French controlled the region but had not explored westward before Jean Nicolet landed near Green Bay in 1634. Their period of exploration, missionary efforts, and fur trading lasted until it was ended by the British capture of Quebec in 1760 and the Treaty of Paris in 1763, which ceded the region to the English.[20]

HABITAT TYPES

Pre-European forests consisted of large tracts of sugar maple (*Acer saccharum*), American beech (*Fagus grandifola*), eastern hemlock (*Tsuga canadensis*), yellow birch (*Betula alleghaniensis*), and red and white pine (*Pinus resinosa* and *Pinus strobus)*. Fire, which periodically altered this association, allowed early successional species, mainly white pine (*Pinus strobus*), paper birch (*Betula papyrifera*), and aspen (*Populus* spp.), to become abundant. Swamp forests were dominated by black spruce (*Picea mariana*) and balsam fir (*Abies balsamea*).[21]

18-21 Stearns, Forest W., *History of the Lake States Forests: Natural and Human Impacts,* In: Vasievich, J. Michael; Webster, Henry H., eds., Lake States Regional Forest Resources Assessment: Technical Papers. Gen. Tech. Rep. NC-189. St. Paul, MN: U.S. Department of Agriculture, Forest Service, North Central Forest Experiment Station: 8-29, 1997

19 *Wildland Fire Management Plan*, Seney National Wildlife Refuge, Department of Interior, Fish and Wildlife Service, 2008

During the late 19th century, the pine forests and, to a lesser extent, the hardwood stands were heavily logged. On the other hand, the swamp conifers were logged only to a limited extent, as access to this resource was difficult. Early logging was usually associated with uncontrolled fires that burned much of the soil humus needed for rapid forest regeneration. As a result, much of the forest regrowth and maturation (i.e., the closing of forest openings) has occurred only in recent years. Large areas of present vegetation consist of the aspen-birch and white-red-jack pine types.

Due to many factors, the present condition of the second-growth forest is quite variable. There is an abundance of sugar maple, beech, and yellow birch on good upland sites. Swamp forests presently cover large areas of wilderness.

Today's four major land cover types in the Seney NWR are shrub/scrub, which covers 26% of the Refuge; coniferous forest, which covers 24%; marsh-bog, covering 14% and broadleaf forest, acres covering 6%.[22] The remainder is mostly wetland forests, open lands, and open water.

LOGGING HISTORY

The township and section lines were vital for any logging operation; without them, it was impossible to establish proper ownership. William Burt, a noted surveyor, established most of the exterior township lines of the Upper Peninsula in 1841. From 1850 to 1855, William Ives and George Adair established the interior lines.[23] When the first surveyors walked the eastern Upper Peninsula, the primary forest tree species were sugar maple, beech, yellow birch, and eastern hemlock (*Tsuga canadensis*). Swamps were generally cedar, black spruce, and eastern hemlock. But it was the stands of large white and red pine

22 *Wildland Fire Management Plan*, Seney National Wildlife Refuge, Department of Interior, Fish and Wildlife Service, 2008

23 National Park Service, U.S. Department of the Interior, Pictured Rocks National Lakeshore, *Logging History*, https://www.nps.gov/piro/learn/historyculture/logging-history.htm, Last updated: November 17, 2021, accessed May 4, 2022

that attracted the rugged early lumberjacks.

The commercial lumbering of Michigan began in the 1840s when it was growing obvious to eastern logging companies that they were coming to the end of their forests. Owners, investors, and a few criminals started to visit Michigan, the next untapped area of the great pine belt. The land owned by the Federal government was granted to Michigan. Once the State properly surveyed it, land grant offices began selling government land in "forties," or 40-acre lots. Land sold for about $1.25 an acre in 1850.[24]

When the first "land lookers" and timber cruisers arrived, the Upper Peninsula was a vast forest of white pine. No significant highways existed in the area. In 1882, Alger, Smith & Company began logging in the eastern Upper Peninsula, and other companies soon followed.[25] Like any other enterprise, the logging industry depended on transportation for moving products and personnel, and the town of Seney secured its place in history through the railroad. The town's name came from one of the major investors in the railroad, George Ingraham Seney, a New York banker who had invested his own money.[26] Seney was suddenly one of the centers of the logging industry and grew abruptly to become a thriving town of around 3000 people and many more when the lumberjacks came in from the 15 or more nearby camps. There were about twenty-one saloons, ten hotels, various businesses, several hundred houses, and more than a few illicit diversions for the lumberjacks.

When the lumberjacks and river men came to town, there was more

---

24 Loomis, Bill, *Shanty Boys, River Hogs and the Forests of Michigan*, Business Section, The Detroit News, Apr 11, 2012. http://blogs.detroitnews.com/history/2012/04/11/shanty-boys-river-hogs-and-the-forests-of-michigan/

25 *A Little History of Seney, Michigan in the Upper Peninsula*, http://www.exploringthenorth.com/seneymi/history.html

26 Jobst, Jack, *Hemingway in Seney,* Michigan Historical Magazine, November/December 1990, https://www.cmich.edu/library/clarke/ResearchResources/Michigan_Material_Statewide/Hemingway_in_Michigan/Pages/Hemingway-in-Seney.aspx

crime and lawlessness than any other small town in the Midwest. They worked hard, and they played even harder. Feisty, fearless lumberjacks weary of woods, wilderness, and a lack of women came to town to enjoy life and have some fun. Rough and crude, the "jacks" drank, caroused, and fought, and many a bout of boasting, gambling, and fisticuffs turned sour, and blood was spilled and was sometimes fatal.[27] The quiet little village of Seney became notorious. One reporter wrote that no one could "see how any place in the pineries could have come closer to hell than Seney."

Logging was conducted during the winter as moving the logs on sleds or with 10-foot-tall "big wheels" was easier because of the snow. Oxen, mules, or horses pulled the sleds and "big wheels." The logs were hauled to the Fox and Manistique Rivers and piled up until the spring thaw, and then they traveled down the river to Manistique, where they were sent to the sawmills. This wood was used to build homes throughout the Midwest and prairie states.

Logging operations forever altered the landscape of the Upper Peninsula's great forests. The lumberman's ax echoed through the woods as local mills depleted the region's valuable red and white pine supply. After the pine forests were cut, mill owners turned their axes and saws to the Refuge's northern hardwood and swamp conifer communities.

Following the logging operations, fires were often set to clear away the debris. Some of these fires burned deep into the organic soil and killed the seeds that would have produced a new forest. On many Refuge areas, the scars from these logging operations remain visible to this day.

---

27 Johnson, Coralie Cederna, *A Shadowy Saga of Seney*, reprinted from Peninsula People Magazine, https://wildwoodpress.org/a-shadowy-saga-of-seney/, 2017

CHAPTER TWO

THE REFUGE

REFUGE HISTORY

Native American use of the pre-European landscape was minimal; the first significant changes to the landscape started with logging the mixed-pine forests in the late 1800s and early 1900s; timber operations forever changed the landscape of the Upper Peninsula's great pine forests. Once the old-growth pine was gone, the mill owners turned their axes and saws to the northern hardwood and lowland coniferous forests. Fires were often set after logging to clear away the debris and sometimes burned deeply into the soil, damaging its quality and killing the seeds that would have produced a new forest.

Around 1908 the Western Land Surety Company, a land syndicate, bought the Seney marshes from the Cleveland Cliffs Mining Company. The company paid approximately one dollar per acre, and from 1912 to 1915, their development program dug several drainage ditches in the marsh. The Walsh Ditch is sixteen miles long, beginning north of highway M-28 and eventually draining south into the Manistique River; they planned to drain the area and promote it

Figure 5. Seney National Wildlife Refuge Sign

as prime agricultural land (Appendix, page 211).[28]

Through propaganda and extravagant promises of the rich-looking black peat soil's agricultural potential, farmers bought land, frequently sight unseen, with plans of establishing homesteads. They soon found that the black fertile-looking peat was underlain with a coarse porous sand base, and as soon as the spring rains stopped, newly planted crops quickly wilted and dried, and the peat soil was very susceptible to frost. Once the marsh vegetation was turned under by the plow, exposing the peat to the surface, it quickly dried out, and the wind soon blew it off, leaving nothing but the bare sand. One by one, the discouraged farmers gave up, abandoned their homestead holdings, and moved away. The land eventually reverted to the State for tax delinquency.[29]

In 1934, the Michigan Conservation Department recommended to the Federal Government that the Seney area be developed for wildlife. The proposal was accepted, and the Seney National Wildlife Refuge was established in 1935 by President Franklin D. Roosevelt to protect and produce migratory birds and other wildlife (Figure 5). Much of the tax-reverted land owned by Michigan was transferred to the Fish and Wildlife Service. Some holdings were purchased from their owners, and about one-quarter of the total acreage was acquired through the power of "eminent domain."[30]

The new Refuge was intended to be a haven for migrating waterfowl. Physical development and restoration of the Refuge land began soon after its establishment. The Civilian Conservation Corps (CCC) and the Work Projects Administration (WPA) assisted in building an intricate system of dikes, water control structures, canals, and roads.[31]

---

28 Losey, Elizabeth Brown, *Seney National Wildlife Refuge, Its Story*, Lake Superior Press, Marquette, MI, 2003, p. 5

29 Losey, *Seney National Wildlife Refuge, Its Story*

30 Losey, *Seney National Wildlife Refuge, Its Story*, p. 6

31 *Seney National Wildlife Refuge-General Brochure*, https://www.fws.gov/uploadedFiles/Region_3/NWRS/Zone_2/Seney_Complex/Seney/Sections/Newsletters/

This system now impounds 7,456 acres of open water in 27 major pools ranging from approximately 100 to 1000 acres. The principal source of water for these pools is several streams and ditches that flow into the Refuge from the north. Because the general topography of the Refuge is flat with natural drainage to the southeast, water flows from one pool to another without the aid of pumping stations.

The Refuge is not all marsh. There is a rich diversity of habitat types represented within its boundaries. Bottomland timber borders the rivers. Pine-covered sandy knolls and ridges are numerous. Northern leatherleaf bogs occur frequently. Maple, birch, and beech hardwoods are found in a broad zone bordering the Manistique and Driggs rivers and Pine Creek. Open sedge marshes and thick growth of cedar, tamarack, poplar, and spruce are characteristic of the western third of the Refuge. To the north are extensive grasslands along with scattered groves of poplars. These were kept open by the countless wildfires that burned the peat soil right down to the underlying sand. This was once excellent country for upland grouse, such as the greater prairie-chicken (*Tympanuchus cupido)* and sharp-tailed grouse (*Tympanuchus phasianellus*). However, with fire suppression, the open grassland rapidly closed in and reverted to willow brush and poplar growth, dramatically lessening its suitability for the upland grouse.[32]

The Refuge was established when suppression was the answer for dealing with fires on the landscape. In 1936, the staff spent two-person years suppressing marsh and wildfires. In 1937 the Refuge was outfitted with firefighting equipment, and they erected a fire tower. In 1941 the Civilian Conservation Corps cleared and burned materials that could fuel wildfire as a hazard reduction measure.[33]

Sharp-tailed grouse require an open landscape; fire or logging

---

SNYGeneralBrochure.pdf

32 Losey, *Seney National Wildlife Refuge*, Its Story, p. 7

33 Marsh, Dawn S., *The Evolution of Land and Fire Management at Seney National Wildlife Refuge, From Game to Ecosystem Management*, https://www.fws.gov/uploadedFiles/Marsh_2013_fire_conceptual_changes.pdf, p. 5

must set back encroaching vegetation to maintain their habitat. In 1944 the Refuge was met locally with conflicting opinions regarding the value of fire in improving habitat for another species, the greater prairie-chicken. The staff was interested in conducting a prescribed burn; the proposal to burn for the greater prairie-chicken habitat was approved; however, the 11,200-acre prescribed fire was unsuccessful due to heavy rains. In 1946, the Refuge still focused on improving the greater prairie-chicken habitat, as their populations were declining.[34]

Attempts to burn in 1947 and 1948 met with mixed results. Some locations were not fit for burning, and the objectives were unattainable. But the idea of using prescribed fire to improve game habitat had gained support among stakeholders. Eventually, the State agreed to work with the F&WS and cooperatively burn upland game habitat in the fall of 1952 when they attempted to determine the effect of marsh burning on willows, tag alders, and other brush. Still, the project failed due to poor plot placement. The Refuge successfully used fire in 1954 as a land management tool, repeatedly garnering considerable local support.[35]

The Refuge continued to use prescribed fires to promote wildlife habitat well into the later part of the 20th century. In 1970, the Refuge manager wrote that the Refuge used prescribed fires to "kill encroaching woody vegetation, clean the area of thick litter, and stimulate the growth of green succulents for geese to feed on." Wildfires on the Refuge were infrequent and were often started by recreationalists.[36]

The Refuge maintained cooperative agreements with the State over the years to assist each other in fire suppression if needed. Such agreements were in place when the Walsh Ditch Fire of 1976 occurred. In 1976 both agencies were using a shared, low-band radio network.

34-36 Marsh, *The Evolution of Land and Fire Management at Seney NWR*, p. 6

REFUGE LOCATION

The Upper Peninsula can boast about being home to the three oldest cities in the State of Michigan. The oldest city in Michigan, Sault Ste. Marie was founded in 1668. St. Ignace is the second oldest, founded in 1671, and Marquette is the third oldest, founded in 1675. Father Jacques Marquette is credited with establishing Sault Ste. Marie in 1668. Driving from the third oldest city in Michigan (Marquette), you head east on Highway M-28. After you pass through Shingleton, you will find yourself on what many call the "Seney Stretch."[37]

This so-called Seney Stretch is a 24-mile-long section of M-28 that is nearly perfectly straight and level. It is the longest straight stretch of road in the State of Michigan and is believed to be the second-longest straight stretch of road in the Midwest. At the western end of the Seney Stretch is the town of Shingleton, and at the eastern end is the town of Seney. Near the Creighton River, there used to be a Standard gas station, the Pines bar and restaurant, and cabins for Seney Stretch motorists. It was also a popular base camp for hunters and anglers.[38]

Michigan Highway M-77 is at the eastern end of the Seney stretch. If you turn south on M-77 and go about 6 miles, you will find the entrance to the Seney National Wildlife Refuge on the west side of M-77.

The Seney National Wildlife Refuge is in the east-central portion of Michigan's Upper Peninsula, halfway between Lake Superior and Lake Michigan (Figure 6).[39] The Refuge encompasses approximately 95,238 acres; the Seney Wilderness Area and the 9600-acre Strangmoor Bog National Natural Landmark encompass 25,150 acres or 26

37-38 The Mining Journal, Marquette, MI, *Seney Stretch and Seney Wildlife refuge*, July 15, 2014

38 The Mining Journal, Marquette, MI, *Seney Stretch and Seney Wildlife Refuge*, July 15, 2014

39 *U.S. Fish & Wildlife Service, National Wildlife Refuge System*, https://www.fws.gov/refuges/airquality/aris/sene/index.html

Figure 6. Location of the Seney National Wildlife Refuge

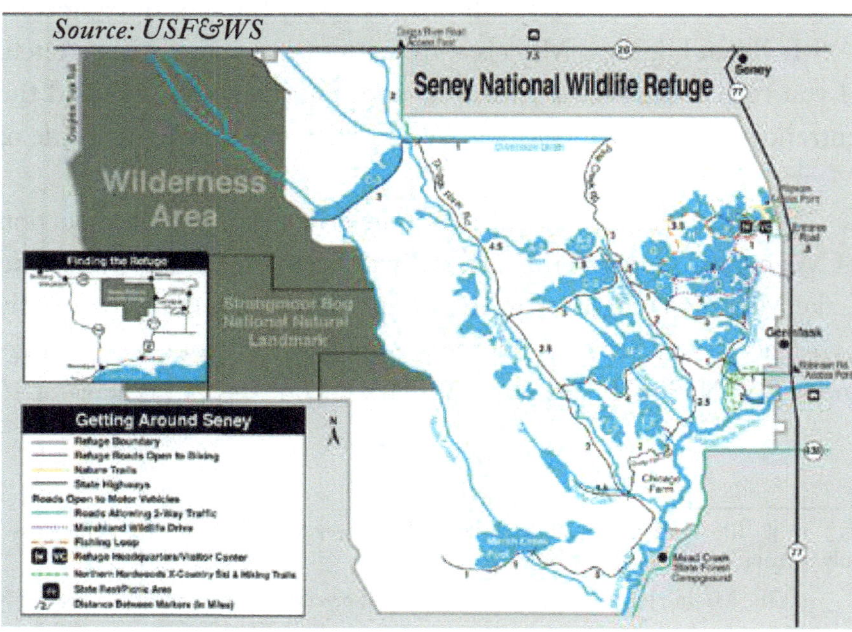

Figure 7. Map of the Seney National Wildlife Refuge

percent of the Refuge (Figure 7).[40] Located in northeastern Schoolcraft County, the Refuge is removed from major population centers; the three nearest significant communities are more than eighty miles away.

CLIMATE

The climate of the Seney NWR is influenced considerably by its proximity to Lakes Superior and Michigan.[41] The most common winds from spring through early fall are from the southwest to northwest, averaging approximately ten m.p.h. The following climate data for the Seney National Wildlife Refuge is from 1961 to 1990.

Temperature: In winter, the average temperature, is 18.1 degrees, Fahrenheit (F), and the average daily minimum temperature is 8.8 degrees (F). The lowest temperature on record was –41 degrees (F) on February 17, 1979. In summer, the average temperature is 64.4 degrees (F), and the average daily maximum temperature in summer is 77.2 degrees (F). The highest temperature on record was 100 degrees (F) on July 30, 1975.

Precipitation: The average annual total precipitation is 31.77 inches; about thirteen inches, or roughly 42 percent, usually falls from June through September. Thunderstorms occur about 29 days yearly, most occurring between June and September. During this period, the heaviest 1-day precipitation amount was 4.40 inches on July 19, 1975.

Snowfall: Average seasonal snowfall is around 121 inches. The most significant snow depth during this period was forty-two inches on March 5, 1977. There is at least one inch of snow depth 129 days per year; the heaviest 1-day snowfall on record was 14.0 inches on March 3, 1985.

---

40 *Seney National Wildlife Refuge-General Brochure,* https://www.fws.gov/uploadedFiles/Region_3/NWRS/Zone_2/Seney_Complex/Seney/Sections/Newsletters/SNY-GeneralBrochure.pdf

41 *Climate Narrative for Schoolcraft County,* Michigan, Prepared by the Natural Resources Conservation Service National Water and Climate Center, Portland, Oregon, www.wcc.nrcs.usda.gov/legacy/ftp/support/climate/soil-nar/mi/schoolcraft.doc

Relative Humidity: The average relative humidity in mid-afternoon is about 55 percent in May and nearly 75 percent in December. Humidity is higher at night, and the average at dawn is about 80 percent in most months, except near 90 percent from June to September.

During the spring and summer months, onshore breezes cause frequent afternoon thunderstorms. Lightning strikes are common during such storms. Growing season evaporation averages 25.1 inches. It is expected that during only 5 percent of the time will drought indexes reach extreme severity levels. The growing season averages 119 days.

CHAPTER THREE

FIRE HISTORY

ANCIENT FIRES

Fire first appeared on Earth long before the coming of humans, but it has not always been a feature of this planet. Combustion requires the presence of oxygen, but it was not an important component of the earth's early atmosphere.

It slowly accumulated as a photosynthetic by-product of cyanobacteria, algae, and, later, plants. Initially, an abundance of reduced iron and other material in Precambrian waters would have absorbed any oxygen released, so it was only about 2.3 billion years ago that free oxygen formed a significant component of the Earth's atmosphere. Another requirement for fire is the presence of combustible fuel, and this, too, was the product of living organisms. The rise of fire and its intimate relationship with plants is a story that has changed our earth for the past 360 million years, but there was a time when fires did not exist. Fossil charcoal indicates that wildfires began soon after the appearance of terrestrial plants. The occurrence of wildfires throughout the history of terrestrial life invites conjecture that fire must have had pronounced evolutionary effects on most ecosystems' flora and fauna. Earth is an intrinsically flammable planet owing to its cover of carbon-rich vegetation, seasonally dry climates, atmospheric oxygen, and widespread lightning and volcanic ignitions.[42]

There are three components needed to start a fire: fuel, oxygen, and heat, commonly referred to as the fire triangle.[43] The fire triangle

---

42 *Wildfire*, Wikipedia, https://en.wikipedia.org/wiki/Wildfire#cite_note-:0-5 3 February 2022

43 U.S. Fish and Wildlife Service, *Role of Fire in Alaska*, https://www.arlis.org/

is a simple way of understanding the three elements a fire needs to ignite. Each side of the triangle represents one of the three ingredients, oxygen, heat, and fuel, demonstrating the interdependence of these ingredients in creating and sustaining fire. A fire occurs when the elements are present and combined in the right mixture. When there is not enough heat generated to sustain the process, when the fuel is exhausted, removed, or isolated, or when the oxygen supply is limited, then a side of the triangle is broken, and the fire will die or not start at all.[44]

Of fire's three essential elements, only the heat of ignition thrived on the early Earth. Oxygen did not begin to collect until the last two billion years and did not approach modern quantities until five hundred million years ago. Land plants suitable to carry combustion did not become abundant until four hundred million years ago. The onset of frequent and widespread occurrences of charcoal (a proxy for past fires) 360 million years ago, is eighty million years after terrestrial plants began to colonize earth about 440 million years ago. The Devonian (419-359 million years ago) witnessed a rapid rise in vegetation, with woody shrubs and small trees dominating the landscape. During the late Devonian, Archaeopteris, a woody conifer-like tree genus, began to develop. By the end of the Devonian, there was a rapid expansion of woody trees, able to support their structure to stand upright while delivering nutrients throughout. During this rapid expansion of woody trees, photosynthesis changed the atmospheric composition. This sudden rise in photosynthesis pulled carbon dioxide from the atmosphere and replaced it with oxygen. Atmospheric oxygen in the middle to late Devonian is thought to have increased oxygen from around 13% to 17% (oxygen is at 21% today).

Forests developed through the Devonian, increasing the atmospheric

---

docs/vol1/83599936/firetriangle.html, US F&WS Alaska Region Division of Information Resource Management (IRM) and External Affairs (EA)

44 *Wildfire 101: The Fire Triangle and the Fire Tetrahedron*, https://www.redzone.co/2016/02/17/wildfire-101-the-fire-triangle-and-the-fire-tetrahedron/, 17 February 2016

oxygen to a key level, allowing fires to readily burn using the available oxygen and carbon to create carbon dioxide. The rise of forestation during the Devonian led to the potential for wide-ranging fires across terrestrial forests. This period defined a new regime in earth's terrestrial ecosystems, where worldwide fires utilized the abundance of fuel in the form of woody vegetation and the newly formed oxygen.

The earliest known smoldering wildfire occurred approximately 419 million years ago during the Silurian period. A team of scientists from Cardiff University's School of Earth and Ocean Sciences (UK) found evidence of this low-intensity burn, started by a lightning strike, while studying three-dimensional charred fossils of small plants found in rocks near Ludlow in the Welsh Borderland, UK in 2004.[45]

These earliest fires would have resulted from combustion caused by lightning or the heat of volcanic action. Later fires awaited the appearance of the hominids. Imagine those early humans, at the mercy of powerful predators, gathering burning sticks from a lightning strike to bring back to their campsite for protection during a long night of fitful sleep, and finding their fire had gone out with no way to reignite it. Prehistoric hunter–gatherers used fire to drive game, cook their food, to keep predators from using dense vegetation around their campsites as cover to ambush them, and possibly as a weapon against each other. Human-caused fire has existed in some form for most of the Pleistocene, two million years or more, though it became a growing planetary presence across the Holocene during the last 10,000 years. But it was when people took to agriculture that fire became a necessity. It cleared the way for the planting and harvesting of crops; it complemented and competed with nature's fire, and this changed the nature of the world's habitats forever.[46]

45 *First Wildfire*, https://www.guinnessworldrecords.com/world-records/first-wildfire#:~:text=The%20earliest%20wildfire%20smouldered%20approximately,have%20been%20higher%20than%20today's,2022

46 Brain, C., *The three ages of fire*, Nature, 414, 584, https://doi.org/10.1038/414584a, 2001

FIRE IN THE SENEY NWR

The Seney NWR includes glacial outwash channels and a matrix of patterned fens interspersed by sand ridges. Red pine forest ecosystems historically dominated the Seney NWR, along with eastern white pine and jack pine (*Pinus banksiana*). The landscape now contains more jack pine and fire-sensitive species, including American beech, sugar maple, and red maple, along with other coniferous species.

In 2011 researchers reported on the fire history of this landscape during three significant periods that shaped the mixed-pine systems in the eastern part of the Upper Peninsula. The periods studied were pre-Euro American settlement, defined as before 1860. Then from about 1860 to 1935, a utilization period during which there was Euro-American settlement going on, along with the exploitation of the forest resources. Then there were many slash fires, followed by attempts to convert the land to agriculture, which typically failed. Many of these parcels were abandoned, and the ownership reverted first to the State and then to the Federal Government.

Mixed red and white pine forests in eastern Upper Michigan saw frequent fires, about every 50–60 years, before Euro-American settlement. Post-settlement, the fire cycle lengthened, and forest composition has shifted to include more jack pine and fire-sensitive deciduous trees, increasing fuel loadings and changing wildlife habitat.

Researchers found that historically, fires in these landscapes were not stand-replacing. These systems were maintained by low-intensity fires or maintenance fires that created forest stand structures that are much different than the present. Before European settlement, the research revealed that low-severity fires burned about 12,000 to 25,000 acres in this area every 50 to 60 years, which was more frequent than expected. Forest management since settlement has lengthened that natural fire return interval, increasing fuel loads and shifting forest composition away from mixed pines to maple and aspen. Around the turn of the 20th century, the forests were high-graded (only the largest and most valuable trees were cut), leaving large swaths of single-aged trees. If a fire did start in many of these areas today, it might be stand-replacing. The fuel loadings in some stands are currently reported at potentially dangerous levels.[47]

Landforms influence fire history, forest composition, and structure. Tree ring analysis showed that the glacial outwash channels and sand ridges at Seney historically burned on average every 53 years and 47 years, respectively. Many of these fires were small and isolated. Still, larger landscape-scale fires occurred every 160 years for mixed jack, white, and red pine forests versus about 320 years for eastern white pine and red pine-dominated tracts. These larger fires also appear strongly associated with large-scale regional and continental climatic patterns.[48]

Fire history differed between two different landforms that dominate this glacial landscape. There are embedded sand ridges that are wind-blown glacial deposits. In the eastern Upper Michigan, these ridges may be 350 feet across by up to three-quarters of a mile long, surrounded by wetlands, and are not connected. The tree ring analysis showed that the intervening fens would dry out and burn when a widespread drought occurred, and the sand ridges would become connected for short periods.[49]

The other landform of particular concern was outwash channels. These are very sandy areas that occur along streams and rivers. They were channels underneath the glaciers 10,000 years ago. They tend to be linear, narrow, and more connected than sand ridges. Tree ring analysis revealed that fires more significant than 250 acres in this landform resulted from widespread drought, as described above. Many of the red pine here got their start after fires that occurred across the

---

47-50 Goebel, P. Charles, et. al., *Fire History Study Reveals Surprises about Mixed-Pine Ecology in Eastern Upper Michigan*, Fire Science Brief Issue 132 April 2011 Page 6 JFSP Project Number 05-2-1-86, https://www.firescience.gov/projects/briefs/05-2-1-86_FSBrief132.pdf

sand ridges disbursed throughout the Seney swamp.[50]

Researchers, starting over 75 years ago, began investigating forest fires throughout history and their effects in the Upper Peninsula. They all point out that these ecological disturbances have been extensive in the area. Fire history research at the Seney NWR by staff and university scientists found evidence of forest fires as early as 1596. They discovered six landscape-scale fires in the Seney NWR of ten thousand acres or more over the last 350 years.[51]

Large fires occurred in 1754, 1791, 1864, 1891, 1910, and 1976. They determined that the mean fire return interval for large fire events was 37 years, with a range of 19-73 years. These natural fires occurred in the early, mid, and late season, but large fires were late-season events.[52]

Since the beginning of settlement, about 1825, forest fires have occurred annually in the Upper Peninsula of Michigan. Early field notes by surveyors made some mention of fire in the mixed forest-marsh areas. This was particularly true if the site was situated in any sizable area of coarse sandy soils known as Grayling sand. All the present jack pine areas were burned over periodically, in pre-settlement times, since their establishment after the glaciers retreated. The ancient fires that burned the jack pine were unhindered except by rain and natural barriers. The encroachment of most of those fires on swamps and surrounding deciduous forests assisted in maintaining suitable habitat for such species as ruffed grouse, eastern cottontail, deer, and bear. Large areas in northern Michigan were desirable deer range for hundreds of years because of the effects of fire.

Before European settlement, fires occurred regularly on both the extensive peatlands and on the dunes within the peatlands of the Seney

---

51 Goodrich, Marcia, *Back to Nature,* Michigan Tech Magazine, Spring 2013, Volume 50, Number 1, p 8-13 Michigan Technological University

52 Corace, Greg; Goebel, Charles; Kashian, Dan, Seney National Wildlife Refuge, *Reconciling Yellow Rail Habitat Use and Landscape Dynamics at Seney National Wildlife Refuge,* https://slidetodoc.com/seney-national-wildlife-refuge-reconciling-yellow-rail-habitat/

NWR approximately every 29 years. Thirty-three wildfires occurred on the Refuge from 1936 to 2001, with the average occurrence of a fire every two years. The first wildfire documented in annual Refuge reports took place in 1944 and was caused by lightning.[53] D. Q. Thompson, a fire researcher, investigated the frequency and distribution of lightning fires in the Seney swamp. To determine the probability that lightning would strike and kindle a fire, he investigated fire reports in the Seney NWR files and found records of lightning-caused fires in 1954, 1964, 1967, and 1976. In an unpublished report, Thompson concluded that such fires were not likely to occur annually but within each decade. All the fires originated in remarkably similar locations. Thompson also showed that fire was correlated with periods of low water levels; the severity of the 1976 drought predictably caused difficulty in controlling the Seney wildfire,[54] the last landscape fire in the Refuge.

HISTORICAL FIRES

Following is a brief history of many of the significant forest fires of record in Michigan and particularly the Upper Peninsula. Based on General Land Office notes, fires were a regular disturbance of the native ecological communities of the Upper Peninsula. Most of the larger fires that occurred during the late 19th and early 20th centuries were associated with extensive logging and residual slash.

1863

The little village of Harvey, near Marquette, had a setback in May 1863 when it was almost destroyed by fire. The headline in the Lake Superior Journal read: "Terrible Conflagration-Village of Harvey in Ashes-Great Destruction of Property-One Hundred People Without

53 *Wildland Fire Management Plan, Seney National Wildlife Refuge*, Department of Interior, Fish and Wildlife Service, 2008

54 Anderson, Stanley H., Resource Publication 146 July 1982 *Effects of the 1976 Seney National Wildlife Refuge Wildfire on Wildlife and Wildlife Habitats*, Fish and Wildlife Service U.S. Department of the Interior

Homes." It read: "Carried by a terrible south wind, the fire came suddenly on the town and swept around and through it with hardly any warning, and the scenes in the place were disheartening." Eighteen buildings burned totally, and over one hundred people were left homeless. Two large boarding houses, the schoolhouse, one store, a warehouse with goods, Mr. Harvey's office, and the sawmill with a large amount of lumber were destroyed. The fire had its origin in the clearing of a farmer who resided within a mile of the village. He had been previously warned about burning stumps and trees as other fires had gotten out of control.[55]

1871-1884

One of the first recorded regional fires was indicated in the General Land Office notes on October 8, 1871, when approximately 50,000 acres burned in the southern Upper Peninsula, a northern extension of the Peshtigo Fire that burned an estimated 1,200,000 acres and killed 2,500 people.

Also on October 7, 1871, in the lower peninsula, the Great Michigan Fires burned at multiple locations from Lake Michigan to Lake Huron, consuming 1.5 million acres, killing 500 people, and burning thousands of structures. At the same time, the Port Huron Fire burned 1.2 million acres, killed fifty people, and burned thousands of structures. The Thumb Fire burned one million acres on September 5, 1881, killing 282 people and destroying more than 2,000 structures.[56] Between 1883 and 1885, large fires occurred throughout the central and eastern Upper Peninsula. One very large fire in 1884 was reported to have burned from Wetmore to St. Ignace.[57]

55 Chocolay Township History, *Then and Now*, https://www.chocolay.org/documents/townshiphistory.pdf, June 2008, p. 18

56 *List of Michigan Wildfires*, Wikipedia, https://en.wikipedia.org/wiki/List_of_Michigan_wildfires, 22 January 2021

57 *Eastern Upper Peninsula Regional State Forest Management Plan Section 2, Forest History*, https://www2.dnr.state.mi.us/publications/pdfs/forestslandwater/RSFMP/EUP/EUPSection2.pdf

1896

The summer of 1896 was so dry that the wetlands were no longer wet by summer's end. Small fires burned in area swamps and crept toward Ontonagon. On August 25th, dry southwesterly winds kicked up ahead of strong low pressure to the northwest. The swamp fires outside of town roared in and set the towering piles of lumber and sawdust at the mills on fire. Once the "kindling" caught fire, the town was doomed. Ontonagon was laid to waste in a matter of a few hours. During the height of the holocaust, the town was ablaze in one hundred places at once, and nothing could save it. Three hundred and forty-four buildings burnt to the ground. Among them were four churches, a bank, three hotels, a dozen stores, thirteen saloons, two newspapers, the entire Diamond Match Company plant along with forty million feet of lumber, as well as the barge City of Straits and two iron bridges. The village's courthouse and jail were reduced to ashes along with nearly three hundred residences. In one afternoon, some 2,000 people became homeless. The Ontonagon fire burned the village as well as 228,000 acres in Ontonagon County.[58]

In October, a fire started on the Dead River north of Ishpeming. It burned through to Lake Superior in the vicinity of Big Bay, "denuding the mountains" and covering an estimated 64,000 acres of virgin wilderness. Even Isle Royale was not immune; a substantial blaze occurred there in the vicinity of Chippewa Harbor.[59]

1902

In 1902, a fire occurred near Seney and spread northwest to the town of Munising and north to Grand Marais. This fire is thought to have been particularly intense and destructive. Similar fires occurred the following year and again in 1905 and 1906.

---

58 Bohnak, Karl, *So Cold a Sky: Upper Michigan Weather Stories*, United States: Cold Sky Publishing, 2006, p. 271

59 Leefers, Larry A., Dickman, Donald I., *The Forests of Michigan*, Revised Ed., United States: University of Michigan Press, 2016, p. 162

1906

A series of fires broke out in the dry swamps, fields, and forests of the Upper Peninsula in the spring of 1906 and burned down portions of towns and farming settlements. The fires scorched an area roughly from Newberry in the east to Bessemer in the west and from Lake Superior to the southern border of the peninsula.[60] The Chocolay Valley contained some of the finest farms in Marquette County and was virtually swept clean, as if "visited by a tornado." At least fifty farmers lost their homes and their outbuildings along with cattle and other livestock. North of Marquette, on the way to Big Bay, the town of Birch was saved by fighting fire with fire. A force of men set fire to slashings about three-quarters of a mile from the main sawmill. The wide area burned around the town kept the fire at bay and saved the lumbering community. Farther west, the Houghton County towns of Kenton and Sidnaw were threatened but saved as the wind swung around and directed the flames away from the towns. Near Quinnesec, just east of Iron Mountain, a blaze swept out of control in a field where men were pulling stumps and burning debris. Sixteen families lost their homes, and most of the business district in Quinnesec was destroyed. The dry wind spread the fire quickly; it moved into timber and eventually joined the inferno spreading east toward Escanaba. The fires raged sixty-four miles along the Escanaba and Lake Superior Railroad and then for thirty miles beyond Channing in northern Dickinson County.[61] While no historical record remains of the acreage burned, these fires stand out as among the most widespread and destructive to ever burn across the Upper Peninsula.

1908

In 1908 an incredible total of 449,976 acres were burned in the Upper Peninsula, of which 235,203 acres were slash. Of this total,

60 Bohnak, Karl, *So Cold a Sky: Upper Michigan Weather Stories*, p. 274

61 Bohnak, Karl, *So Cold a Sky: Upper Michigan Weather Stories*, p. 276

130,770 acres were burned in Chippewa County.[62] Also in 1908, a very large fire was reported to have burned from Eckerman to Sault Ste. Marie.[63] Over the next twenty years or so, large fires fueled by the slash left behind by logging occurred several times within the region. [64]

On October 15, the Metz Fire in northeastern lower Michigan burned 300,000 acres with forty-three lives lost, fifteen of them when a rescue train derailed, the rails warped by the intense heat, in a wall of flames. The village of Metz was destroyed by the flames.[65] [66]

1911

The last big Michigan fire in which a community was destroyed occurred on July 11, 1911, when slashing fires in Iosco County wiped out the sawmill towns of Au Sable and Oscoda at the mouth of the Au Sable River, with the loss of a score of lives and damage amounting to over $3,000,000.[67]

EARLY FIRES RECORDED BY THE STATE

Early Michigan Department of Conservation records show there were other significant historical fires in the Upper Peninsula from 1912 through the late 1940s.[68] From 1912 to the mid-1930s, fires were

---

62 Pierce, Charles S., *Eleventh Bicentennial Report of the State Game, Fish, and Forestry Warden*, Winkoop, Hollenbeck Crawford Company, State Printer, Lansing, Michigan, 1909, p. 15

63 *Eastern UP Regional Forest Management Plan, Section 2, Forest History*

64 *Wildland Fire Management Plan, Seney National Wildlife Refuge*, Department of Interior, Fish and Wildlife Service, 2008

65 *Michigan Wildfires*, Wikipedia, https://en.wikipedia.org/wiki/List_of_Michigan_wildfires, 22 January 2021

66 Sodders, Betty, *Michigan on Fir*e, Thunder Bay Press, Holt, MI, 1997, p. 259

67 Mitchell, J.A., and Robson, D., *Forest Fires and Forest Fire Control in Michigan*, Michigan Department of Conservation, In Cooperation with United States Department of Agriculture Forest Service, Lake States Forest Experiment Station, with the University of Minnesota, University Farm, St. Paul, MN, 1950, p. 12

reported to be so common in the eastern Upper Peninsula that smoke hung over much of the area throughout July and August. Many of these notable fires were late-season incidents, impacting settlements and the surrounding forests severely during significant drought periods. During this period, fires in the Upper Peninsula coincided with catastrophic fires in other areas in the great lakes. Some of the most severely impacted sites are still stump barrens today.

1919

Whorl Fire-North of Newberry, acreage unknown, started on the banks of the Two-Hearted River and burned almost to Newberry

1923

Fifty thousand acres burned over in the vicinity of Silver City, Ontonagon County

1925

May 3-Gogebic County, 15,800 acres
May 11-Ontonagon County, 6,500 acres
May 12-Iron County, 7,320 acres
May 15-Alger County, 5,120 acres
May 21-Gogebic County, 12,070 acres
July 19-Schoolcraft County, 16,640 acres
July 22-Alger County, 16,300 acres
August 5-Chippewa County, 5,694 acres

1926

May 5-Dickinson County, 5,280 acres
May 28-Marquette County, 9,520 acres

68 Mitchell, J.A., and Robson, D., *Forest Fires and Forest Fire Control in Michigan*, Michigan Department of Conservation, p.12-14

1930

August 1-September 16, Ontonagon County, 3,058 acres
August 1-September 26, Menominee County, 3,550 acres
August 8-31, Camp 33 Fire, Schoolcraft County, 4,994 acres
August 8-September 12, Mackinac County, 7,490 acres
August 10-September 15, Driggs Fire, Schoolcraft County, 20,936 acres

1931

April 17-25, Iron County, 6,030 acres
August 13-October 7, Ontonagon County, 5,666 acres
August 18-September 25, Duck Creek Fire, Schoolcraft County, 11,130 acres
September 11-October 6, Keweenaw County, 5,320 acres

1933

August 5-September 29, Horseshoe Lake Fire, Delta County, 2,425 acres
August 20-September 7, Pomeroy Lake Fire, Gogebic County, 14,027 acres
August 30-September 7, Bergland Fire, Ontonagon County, 13,252 acres

1936

July 18-August 15, Green School Fire, Schoolcraft County, 9,343 acres
July 20-Luce County, 2,400 acres
July 23-August 28, Isle Royale (now National Park) 36,000 acres
August 1-Marquette County, 1,597 acres
August 8-Echo Lake Fire, Marquette County, 1,702 acres

1942

April 23-West Branch Township Fire, Marquette County, 4,271 acres

1943

October 6-Marquette County, 1,200 acres

1947

August 11-15, Presque Isle River Fire, Gogebic County, 2,115 acres
October 16-Little Girls Point Fire, Gogebic County, 1,704 acres
October 19-Lac La Belle Fire, Keweenaw County, 3,075 acres

1976

The Walsh Ditch Fire, named for the closest landmark, more commonly known as the Seney Fire, started on July 30 from a dry lightning storm. It was spotted by MDNR aerial patrol within the designated wilderness on the Seney NWR the next day. The Refuge Manager thought no action could be taken on the fire since it was in a Congressionally designated "wilderness" area. A chaotic response and inaction allowed the fire to grow to 1200 acres in almost two weeks before any suppression action took place. Several weeks after the initial report, the fire "blew up," spread out of the wilderness and off the Refuge onto State and private land. All told, the fire spread over 72,500 acres during a period of record drought, requiring an interagency firefighting force of more than 1,200 firefighters to achieve containment. It burned until the winter snows extinguished it.

1986

On May 6, "Black Tuesday," as it came to be remembered by many firefighters, a series of more than 100 fires burned in the Upper Peninsula; most of the fires were in Marquette County. It was very windy and dry, and there were predictions of a low-level touch-down of the jet stream; this was confirmed later in the day with winds over sixty miles per hour at times. Most of the fires were caused by down power lines, with a few by arsonists. Probably the most problematic was the Sporley Lake fire burning in jack pine near K.I. Sawyer Air Force Base. The 4,200-acre crown fire caused the evacuation of the 8,000 base residents and the extreme concern of the base commander for the safety of the nuclear weapons storage area. Extreme measures by firefighters, local fire departments, equipment operators, and military personnel successfully kept it from the base. Local fire departments handled most of the fires. Twenty-four of the larger problem fires were handled by the MDNR with assistance from cooperators across the Upper Peninsula. Of these, four fires totaled almost 8,000 acres lost. Several homes, camps, vehicles, and personal property were lost, but it could have been much worse. Radio communications were very difficult with four large fires near one another causing serious confusion and delays.

1987

In the spring of 1987, fires broke out over sections of the Upper Peninsula, the largest resulting from arson. A lack of winter snow and a very warm, dry April led to fire danger that was more critical than the year before. By early May, MDNR statistics showed nearly 250 wildfires, an almost three-fold increase over the same time the year before.

On May 8, fires broke out in Schoolcraft County, north of Manistique, during an especially dry period. Investigators and firefighters on the scene determined that the fires were set by arsonists. The U.S. Forest Service had hand crews standing by. They saw smokes pop up one after another north of town; eventually, they initially attacked six different fires within a relatively short distance on a trail road. The blazes eventually grew into the Eight Mile Road fire. It scorched more than 4,000 acres before being brought under control on May 11. Other smaller blazes brought the total acreage burned to 5,000 acres.

Two hundred firefighters from Maine to Arizona were brought in to suppress the fires. Investigators had been following the trail of the arsonists in the area during the previous three years. Through much hard work following leads and evidence, the perpetrators were caught.

1988

The Stockyard Fire occurred on July 1 near the village of Rapid River. The Upper Peninsula was experiencing near-record drought and fire danger. The fire originated in a mature jack pine stand that had significant mortality from a jack pine budworm infestation. It rapidly developed into a high-intensity, stand-replacing fire. Reported flame lengths in the jack pine stand were thirty feet, with the crowns fully involved. The Stockyard Fire consumed a total of 1,100 acres, causing the closure of US Highway 2, a major travel route through the Upper Peninsula on a holiday weekend. An MDNR firefighter was seriously injured when his tractor-plow was burned over.

1999

The Tower Lake Fire occurred on May 2 in western Marquette County and was first spotted by MDNR aerial patrol. A campfire near the lake was the likely ignition source. Extremely dry conditions and rough, rocky terrain contributed to the difficulty of suppressing the fire. Four hundred people were evacuated, including residents of Champion, a few miles to the northwest. The fire burned more than 5,000 acres before being suppressed.

2007

The Sleeper Lake Fire[69] began from a lightning strike north of the Village of Newberry in Luce County on August 2. It started in the vicinity of Sleeper Lake and burned to the southeast over largely uninhabited marsh and bogs. By August 5, southeasterly winds began

69 *Sleeper Lakes Fire*, Wikipedia, https://en.wikipedia.org/wiki/Sleeper_Lakes_Fire, 19 February 2022

to push the fire toward the north, threatening structures and forcing nearby residents to evacuate. On August 13, the burned area had increased to 18,500 acres, with one building destroyed. Several Federal, State, and local agencies cooperated in fighting the fire and providing emergency services. Suppression operations included 239 personnel, heavy equipment, five helicopters, and a CL-215 water bomber.

2012

The Duck Lake Fire occurred north of the Village of Newberry in Luce County. The fire started with a lightning strike on Wednesday, May 23, 2012, near Duck Lake. It burned 21,135 acres, consuming 136 structures, including forty-nine homes and cabins (including a store and a motel), twenty-three garages, thirty-eight sheds or outbuildings, and twenty-six campers. By June 15, 2012, the fire was 100% contained after almost 43 miles of fire line were built, Three hundred people were involved in the fire response.[70]

The Pine Creek North Fire occurred on the Seney National Wildlife Refuge in Schoolcraft County when lightning ignited wetlands on Monday, May 21. Firefighters used seven amphibious tracked vehicles, two CL-215 air tankers, and a Type-One helicopter to slow and eventually suppress this fire. An effective technique used when dealing with wetland areas of the fire was to compress the fuels by driving over marsh plants with tracked vehicles, pressing them into the water to slow the fire's spread. The fire burned for ten days before being contained, with 3,420 acres burned. The fire burned in fuels like the 1976 Walsh Ditch Fire but without the severe drought conditions.[71]

---

70 *Duck Lake Fire*, Wikipedia, https://en.wikipedia.org/wiki/Duck_Lake_Fire, 16 August 2020

71 *Pine Creek North Wildfire-Seney National Wildlife Refuge*, https: //www.fws.gov/fieldnotes/regmap.cfm?arskey=32575, June 2012

CHAPTER FOUR

FIRE WEATHER

Weather data is collected, and fire danger rating is computed and archived daily during the fire season by resource protection staff, by hand, or through automated systems in all state and Federal agencies. This information identifies areas of concern for fire occurrence and the need for fire suppression readiness, determines staffing levels, predicts potential fire behavior, provides public information, and is archived for historical purposes.

The Forest Fire Division of the Michigan DNR computed fire danger rating in the 1970s using the 1972 National Fire Danger Rating System (NFDRS). In the 1990s, the then Forest Management Division changed to the Canadian Forest Fire Danger Rating System (CFFDRS), a comprehensive system of tools designed to evaluate environmental factors that influence wildland fire's ignition, spread, and behavior.

The CFFDRS is similar to the NFDRS and depends solely on weather readings. The resulting fuel moisture codes and fire behavior indexes are based on a single standard fuel type described as a generalized pine forest, most nearly jack pine and lodgepole pine.[72]

The CFFDRS, like NFDRS, calls for weather observations to be collected from a standard observation site and time with observations to be taken at solar noon when the sun is at its peak directly overhead. In 1976, weather data taken at the MDNR Seney Field Office was collected manually once a day at 1:00 pm EST from their standard observation site. The daily weather data for the 1976 fire season at this

72 Ziel, Robert, *MI Field Guide for CFFDRS Fire Weather Index System*, http://lakestatesfiresci.net/docs/MN_FireWeatherIndex_FieldGuide.pdf, 2014, p. 5

station is found in the Appendix.

In the rest of this book, you will find daily weather information using abbreviations. Weather record abbreviations are Temp=Temperature, Fahrenheit (F); RH=Relative Humidity(%); Wind=Windspeed in miles per hour (a ten-minute average) coming from a cardinal direction.

Fire danger rating, R, refers to the daily fire danger rating computed from the weather data gathered. Adjective ratings are L=Low, M=Moderate, H=High, VH=Very High, and E=Extreme. Refer back to this chapter as needed.

FUEL MOISTURE CODES

There are three fuel moisture categories, or codes, in the CFFDRS, compared to seven found in the NFDRS. These are represented as unitless codes instead of fuel moisture content.[73]

The Fine Fuel Moisture Code (FFMC) represents forest litter fuel moisture under forest canopy shade. It is intended to represent moisture conditions for shaded litter fuels. It ranges from 0 to 101.

The Duff Moisture Code (DMC) represents the fuel moisture of decomposed organic material underneath the litter. It represents moisture conditions for the equivalent of 15-day (or 360 hour-timelag fuels. It is unitless and open-ended.

The Drought Code (DC), much like the Keetch-Byram Drought Index (KBDI), represents drying deep into the soil. It is unitless, with a maximum value of 1000. Extreme drought conditions have produced DC values near 800; at the Seney fire, it approached 600 in September 1976.

FIRE BEHAVIOR INDEXES[74]

There are three fire behavior indexes in the CFFDRS system. The Initial Spread Index (ISI) integrates fuel moisture for fine dead fuels

---

73-74 Ziel, Robert, *MI Field Guide for CFFDRS Fire Weather Index System* and surface windspeed to estimate a spread potential. The ISI is a key input for fire behavior predictions. The ISI is unitless and open-ended.

The Buildup Index (BUI) combines the current DMC and DC to estimate potential heat release in heavier fuels. In Michigan, the BUI is the primary indicator of seasonal severity; it is unitless and open-ended.

The Fire Weather Index (FWI) integrates current ISI and BUI to produce a unitless index of general fire intensity potential. With

INTERPRETATION OF CODES AND INDEXES

These thresholds are used by the Michigan Department of Natural Resources to provide interpretations to FWI observations and forecasts each day. Though they give insight to overall potential across landscapes and areas of responsibility, they need to be combined with specific fuels and terrain for the site to produce fire behavior interpretations when evaluating a specific fire situation.

FFMC	75	Surface fire – spreads some
	80	Surface fire – spreads more continuous
	90	High probability of spot fires
	91	Onset of extreme FB
DMC	20	Duff available for ignition
	30	Fires moderately intense
	50	Extreme fire behavior
	150	Loss of most available moisture
DC	15	Deep organics near saturation
	300	Onset of smouldering combustion
	500	Sustained deep smoldering
ISI	0	No fire spread
	10	Threshold for crowning
	20	Extreme fire behavior
	50	Rarely exceeds this value
BUI	30	Deeper, heavier fuels start to become involved
	60	Potential for increased mop up, safety concerns
	90	Many fires have potential to escape IA
	100	Bad things can happen
FWI	3	Fire spreads
	25	Onset of potentially extreme fire behavior
	50+	Disaster fires

Figure 8. CFFDRS threshold levels used by the MDNR

dry fuel conditions, the FWI is an indicator of extreme fire behavior potential, it is unitless and open-ended.

Today the Michigan Department of Natural Resources Forest Resources Division (FRD) uses the CFFDRS to determine staffing levels. In the spring, the Fine Fuel Moisture Code (FFMC), which refers to the drying of the finer fuels such as leaves and grass, is used along with the Initial Spread Index (ISI), which is wind and FFMC combined, to provide a rating on how easily a fire starts and spreads. The higher

SPRINGTIME ADJECTIVE RATINGS

Begin with Table 1. Using **Fine Fuel Moisture Code** (FFMC) and **Initial Spread Index** (ISI) calculated for today, find the appropriate adjective class. If instructed to use Table 2, determine the adjective rating using the **Fire Weather Index** (FWI) and **Fine Fuel Moisture Code** (FFMC).

TABLE 1	ISI = 0.0 - 1.9	ISI = 2.0 - 3.9	ISI = 4.0 - 7.9	ISI = 8.0+
FFMC = 0.0 to 74.9	LOW	MODERATE	HIGH	SEE TABLE 2
FFMC = 75.0 to 84.9	MODERATE	MODERATE	HIGH	SEE TABLE 2
FFMC = 85.0+		HIGH	HIGH	SEE TABLE 2

TABLE 2	FFMC < 92	FFMC = 92+
FWI less than 30.0	VERY HIGH	VERY HIGH
FWI = 30.0+	VERY HIGH	EXTREME

SUMMERTIME ADJECTIVE RATINGS

Generally, begin using the summer criteria on June 1. Begin with Table 3. Using **Buildup Index** (BUI) and **Initial Spread Index** (ISI), find the appropriate adjective class. If instructed to use Table 4, determine adjective class using the **Fire Weather Index** (FWI) and **Fine Fuel Moisture Code** (FFMC).

<table>
<tr><th>TABLE 3</th><th>ISI 0.0 - 1.9</th><th>ISI 2 - 2.9</th><th>ISI 3 - 3.9</th><th>ISI 4.0 - 5.9</th><th>ISI 6.0 - 7.9</th><th>ISI 8.0 - 9.9</th><th>ISI 10.0+</th></tr>
<tr><td>BUI=0.0 to 44.9</td><td colspan="3">LOW</td><td colspan="2">MODERATE</td><td colspan="2">HIGH</td></tr>
<tr><td>BUI=45.0 to 69.9</td><td colspan="2">LOW</td><td colspan="2">MODERATE</td><td colspan="2">HIGH</td><td>VERY HIGH</td></tr>
<tr><td>BUI=70.0-99.9</td><td>LOW</td><td colspan="2">MODERATE</td><td colspan="2">HIGH</td><td colspan="2">SEE TABLE 4</td></tr>
<tr><td>BUI=100.0+</td><td colspan="7">USE SPRING CRITERIA</td></tr>
</table>

TABLE 4	FFMC < 92	FFMC = 92+
FWI less than 35	VERY HIGH	VERY HIGH
FWI = 35.0+	VERY HIGH	EXTREME

Figure 9. MDNR Adjective Class Matrix

those numbers, the more quickly a fire will start and spread. An interpretation of thresholds for the individual codes is shown in Figure 8.[75] In the summer, they utilize the Buildup Index (BUI) combined with the ISI for staffing guidance. They use CFFDRS codes to determine the Adjective Class Matrix used for staffing guidance shown in Figure 9.[76]

DROUGHT

The 1976 spring fire season in the Upper Peninsula was typical and relatively unremarkable. With the coming of summer, the Upper

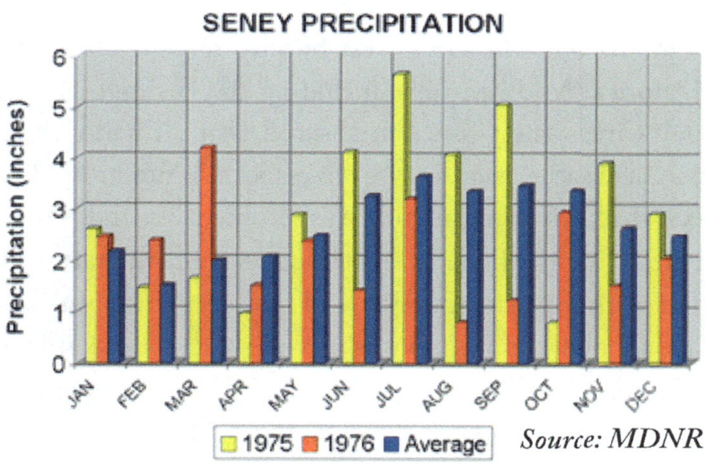

Source: MDNR

- **Average**: 32.19 inches/yr **1975**: 35.91 inches;
- **1976 (whole year):** 25.84 inches (6.35 below normal, 80%)
- **1976 (Jan-July):** 17.41 inches (0.40 above normal, 102%)
- **1976 (June-Sept):** 6.58 inches (7.05 below normal, 48%)

Figure 10. Seney Precipitation 1975-1976

Peninsula entered a period of protracted summer drought.

Data personally gathered from the Marquette National Weather Service (NWS) station showed a drastic departure from historical conditions. From 1872, the stations beginning, there were no drier

75-72 Ziel, Robert, *MI Field Guide for CFFDRS Fire Weather Index System*, p. 7

76 Ziel, Robert, *MI Field Guide for CFFDRS Fire Weather Index System*, p. 13

summers on record prior to 1976. The average summer rainfall for June, July, August, and September at the Marquette NWS is 12.79 inches. In 1976 only 4.79 inches were recorded, a deficit of eight inches, with only one-half of an inch falling in August. Similar readings were reported throughout the UP.

The Seney MDNR (Figure 10)[77] station receives an average of 32.19 inches of precipitation annually. In 1976, there were 25.84 inches, 6.35 inches below average for the entire year. From January to July 1976, 17.41 inches of precipitation fell, slightly above average; from June through September 1976, 6.58 inches fell, 7.05 inches below normal. From July 31 to September 13, only 0.58 inches of rain fell.

In a 1969 study, the US Forest Service researchers investigated seven historical Midwest fires, including the Peshtigo, Thumb, and Hinckley Fires. Their research showed that all had reduced precipitation at weather stations in the respective fire areas. In several instances, the lowest precipitation was recorded over the 3-to-8-month period preceding the fires. When the effects of temperature and precipitation were combined, they found that vegetation was below the wilt stage for one to three months preceding all these conflagrations. Similar drought conditions contributed to the severe fire phenomena by increasing the amount of available forest fuel at Seney in 1976.[78]

EFFECTS OF DROUGHT ON GROUND FIRES

The protracted extreme drought during the summer and fall of 1976 allowed the fire to burn deeply in the Seney swamp, often entirely consuming the upper organic layer of soil. The resulting "ground fire" literally means the ground is burning. While not as spectacular as fast-moving surface or crown fires, they can do long-term damage to the forest and can be extremely difficult and costly to extinguish.

77 Ziel, Robert, Michigan DNR, Incident Command Center, Marquette, MI

78 Haines, Donald A., Sando, Rodney W., *Climatic Conditions Preceding Historically Great Fires in the North Central Region*, North Central Forest Experiment Station Forest Service U.S. Department of Agriculture, Research Paper NC-34, 1969.

Fires of this type may leave nothing but bare rocks with a thin skin of mineral soil in their wake.

Ground fires, once started, are very persistent in drought conditions since combustion needs little or no outside oxygen from the atmosphere. The materials provide such good insulation that very little heat of combustion is lost. Ground fuels develop a water-repellant nature at very low moisture contents and resist wetting.

Fire will persist long and can travel underground in dead roots, especially in large decayed lateral roots extending from snags. When these fires reach the surface, they often start fires outside the control line. Fire can escape control by following root tunnels under the control lines or by long-delayed ignition of materials still unburned outside the line after mop-up is thought to be complete.

The moisture content of the upper soil, as well as that of the covering layer of duff, has an important effect on the fire suppression effort in forest and wildland areas.

Fires in deep organic fuels that have been subjected to long-term drying are of particular concern to fire management staff. The acute effects of drought are not confined to deep organic soils. Dried-out organic materials are frequently embedded in the shallow upper layers of mineral soils. These fuel pockets can decide whether fire lines will hold and become a further problem in mop-up operations. Of course, these were significant considerations on the Seney fire.

When forest fires are human-caused, they are usually found near roads, but when lightning strikes in the middle of a vast swamp, the firefighter's job is compounded. They must carry all their equipment and supplies to the fire, as they had to do on the Seney fire.

When the roots of large trees burn through, the trees may come crashing down at any time in a slight wind, often leaving a trail of fire across the fireline. Finding water to quench this type of fire is also a problem, particularly during a drought. Fighting a fire under these circumstances means carrying in or pumping water long distances. Snags burning many feet above the ground can also be exceedingly challenging to find and extinguish. Ground fires occurring in organic soils during a drought are difficult and expensive to extinguish.

THE PALMER DROUGHT SEVERITY INDEX

The Palmer Drought Index measures dryness based on recent precipitation and temperature.

Month	Rating	Range
June 1976	Midrange Drought	-1.99 to +1.99
July 1976	Midrange Drought	-1.99 to +1.99
August 1976	Moderate Drought	-2.00 to -2.99
September 1976	Severe Drought	-3.00 to -3.99
October 1976	Extreme Drought	-4.00 and below
November 1976	Extreme Drought	-4.00 and below
December 1976	Extreme Drought	-4.00 and below
January 1977	Extreme Drought	-4.00 and below
March 1977	Moderately Moist	+2.00 to +2.99
May 1977	Midrange Drought	-1.99 to +1.99

Figure 11. Palmer Drought Severity Indexes, Seney 1976

In the eastern Upper Peninsula, the drought's effects carried through the winter and gradually returned to midrange drought readings the following May (Figure 11).

Palmer Drought Index records from NOAA, National Centers

for Environmental Information[79] (See Appendix) show a graphic representation of the onset and increase in severity of the drought in the midwest and the eastern Upper Peninsula of Michigan in 1976. In June, the Upper Peninsula showed drought conditions to be mid-range and started climbing rapidly by July. In August, it was -2.00 to -2.99, moderate drought; in September, it was -3.00 to -3.99, severe drought; and from October to December, it was -4.00 and below, extreme drought.

Conditions at Seney were comparable to the historical fires at Peshtigo (Wisconsin) and Thumb Fires (Lower Michigan) in 1881 that burned millions of acres and killed thousands of people. US Forest Service researchers determined that the Palmer Drought Index reached -3.79 for the Peshtigo Fire with similar readings for the Great Chicago Fire, both starting on the same day, October 8, 1871. Their data showed the Palmer Drought Index at Thornville, Michigan, near the Thumb Fire of 1881, was -4.93. Lansing, Michigan, near the epicenter of the lower Michigan Fire of 1871 that burned an estimated 2.5 million acres, had a Palmer Drought Index of -3.04.[80]

The progression and deepening of the drought, as evidenced by the Palmer Drought Index, is shown in the monthly NOAA drought maps for the period April-November 1976 (See Appendix). The drought didn't abate until the following spring.

The Upper Peninsula and the Seney swamp was subjected to abnormal drought conditions in 1976. Recognition and readiness will be the keys to preventing similar disasters in the future. Fire managers must recognize that peat and muck are special wildland fuels whose burning characteristics differ from other natural fuels and must be appraised and monitored differently. They also need to maintain the

---

79 *Historical Palmer Drought Indices*, NOAA-National Centers for Environmental Information, https://www.ncdc.noaa.gov/temp-and-precip/drought/historical-palmers/maps/psi/197605-197704

80 Haines, Donald A., Johnson, Von J., and Main, William A., *An Assessment of Three Measures of Long-Term Moisture Deficiency Before Critical Fire Periods*, https://www.nrs.fs.fed.us/pubs/rp/rp_nc131.pdf

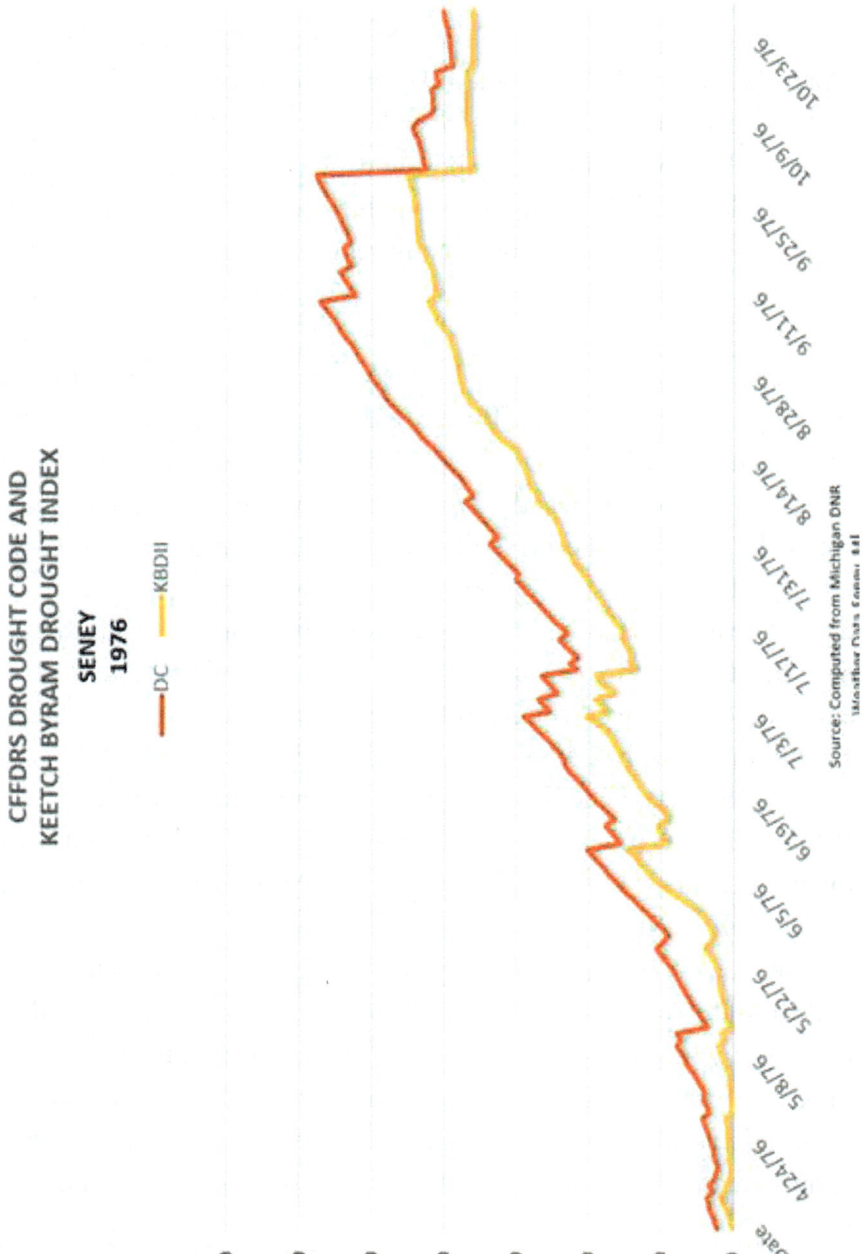

Figure 12. Drought Code and Keetch-Byram Drought Index

vigilance and readiness of a well-trained and equipped firefighting force, a necessity to prevent future conflagrations.

KEETCH-BYRAM DROUGHT INDEX

The Keetch-Byram Drought Index (KBDI) is a stand-alone index that can measure the effects of seasonal drought on fire potential. This index is not an output of the CFFDRS but, in 1976, was calculated by the MDNR-FRD as important information to assess drought conditions. The index's actual numeric value estimates the amount of precipitation (in 100ths of inches) needed to bring the soil back to saturation (a value of zero is complete saturation of the soil). Since the index only deals with the top eight inches of the soil profile, the maximum KBDI value of 800 or 8.00 inches of precipitation would be needed to bring the soil back to saturation. The Keetch-Byram Drought Index's relationship to fire danger is that as the index value increases, the vegetation is subjected to increased stress due to moisture deficiency. At higher values, desiccation occurs, and live plant material is added to the dead fuel loading on the site. Also, an increasing portion of the duff/litter layer becomes available fuel at higher index values.

In 1976, before MDNR-FRD used the CFFDRS Drought Code, it provided MDNR resource protection staff with a good estimate of the severity of drought conditions. The following levels of the KBDI give an interpretation of potential fuel moisture, fire behavior, and mop-up difficulty.[81]

At KBDI Levels 0-150, surface fuels and the ground are moist; heavy fuels are too wet to ignite. Extensive mop up on fires is generally not required. For KBDI Levels 150-300, fuels resist deep-burning, the moisture content of heavy fuels is usually high, and fire behavior is predictable. As KBDI levels exceed 200 and approach 300, more intense burning occurs, and heavy fuels start to be a mop up problem. When the KBDI Levels reach 300-500, fires readily consume surface

81 Melton, Mike, *The Keetch-Byram Drought Index: A Guide to Fire Conditions and Suppression Problems*, Fire Management Notes; Vol 50, No 4, 1989, p. 30-33

fuels with a significant loss of organic matter on and below the surface. Heavier fuels ignite readily and contribute to fire intensity. Escaped fires are difficult to control because heavier fuels contribute to fire intensity. Humidity recovery is not sufficient to extinguish heavier fuels. Organic material will burn into the ground, and mop-up is difficult and time-consuming.

At KBDI Levels greater than 500, heavy fuels readily ignite, including organic soils at depth, frequently burning completely, contributing to abnormal fire behavior. As KBDI levels increase with protracted drought, previously burned organic soils dry out and become available to burn. Roots act as fuses, burning under fire control lines. Fires can burn actively during the day and smolder at depth at night: humidity recovery will not extinguish heavier fuels. Swamps that were full of water in the spring are dry and available to burn.

The DC and KBDI, not surprisingly, followed nearly parallel paths at Seney in 1976 (Figure 12). The drought conditions were very high when the fire started on July 30; the DC was 298, and the KBDI was 207. The maximum reached at Seney in 1976 for the DC was 576 on October 4; the KBDI peaked at 448 on October 3. They both dropped quickly with the cold fall rains that set in just after they peaked. For a historical comparison, the KBDI in 1871 was 300 for the Peshtigo Fire and 381 for the Lower Michigan Fire; it was 435 for the Thumb Fire in 1881.[82]

TOTAL SEASON VIEW-1976

The Fine Fuel Moisture Code (FFMC) represents the fuel moisture of forest litter fuels under the shade of a forest canopy. It is intended to represent the moisture of 16-hour timelag fuels, analogous to fuels approximately one inch in diameter or less. The Initial Spread Index integrates fuel moisture for fine dead fuels and surface wind speed

82 Haines, *An Assessment of Three Measures of Long-Term Moisture Deficiency Before Critical Fire Periods*

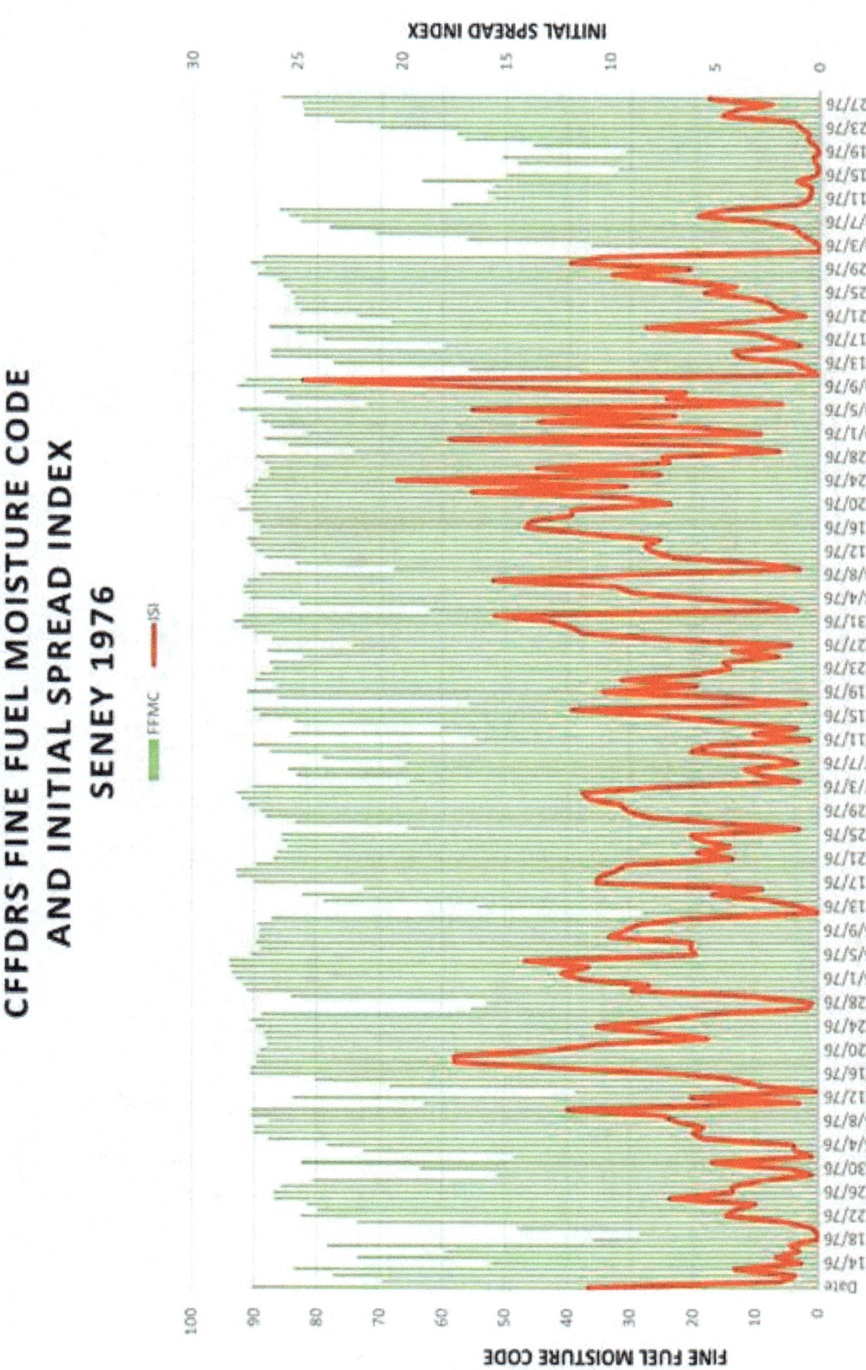

Figure 13. Fine Fuel Moisture Code and Initial Spread Index

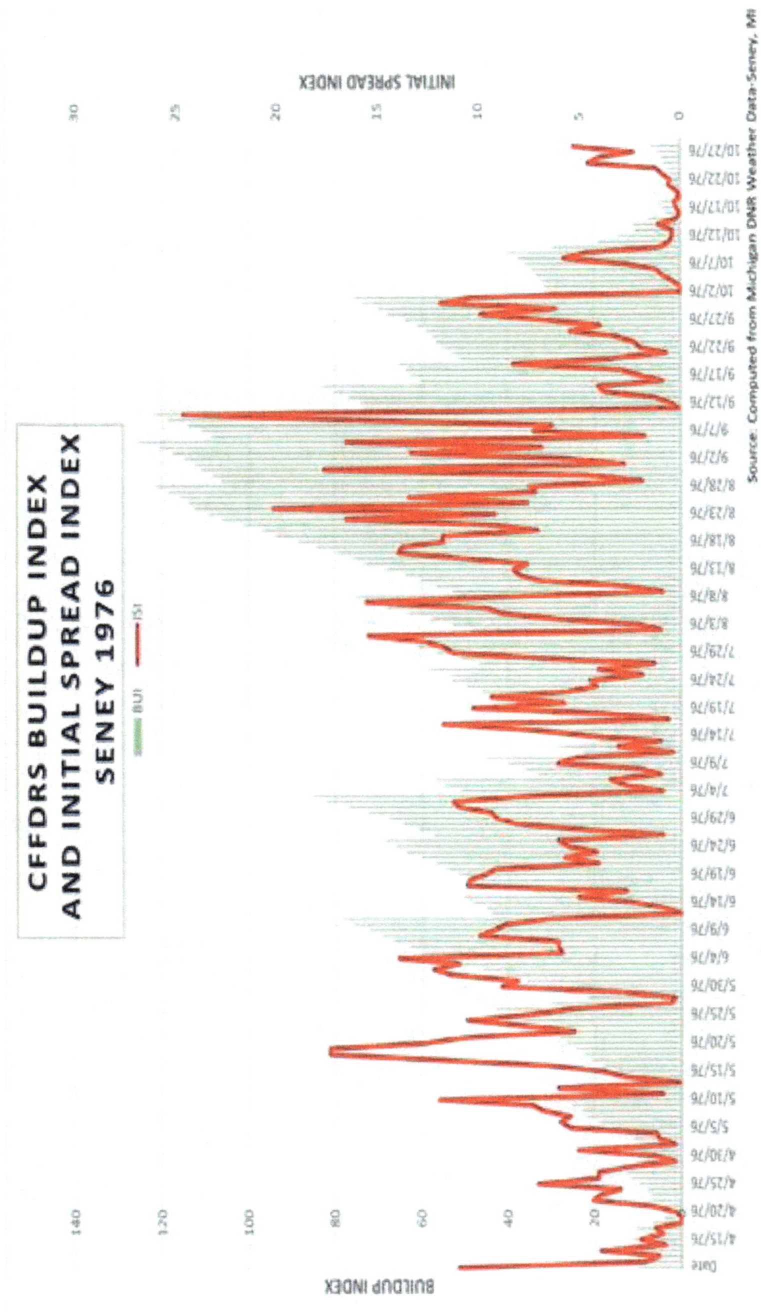

Figure 14. Buildup Index and Initial Spread Index

to estimate the spread potential of a fire. For a view of the Fine Fuel Moisture Code and Initial Spread Index for the 1976 fire season at Seney, see Figure 13. Refer to Figure 9 for adjective ratings.

FIRE BEHAVIOR PREDICTIONS

Two fire behavior indexes, the Buildup Index (BUI) and Initial Spread Index (ISI), are shown in Figure 14. The ISI is described above. The BUI combines the current DMC and DC to produce an estimate of potential heat release in heavier fuels.

<table>
<tr><th colspan="6">INPUT DATA AND CFFDRS OUTPUTS</th></tr>
<tr><td colspan="3">WEATHER DATA</td><td colspan="3">R=FIRE DANGER RATING</td></tr>
<tr><td colspan="3">TEMP=Temperature (F)</td><td colspan="2">L=Low</td><td>M=Moderate</td></tr>
<tr><td colspan="3">RH=Relative Humidity</td><td colspan="2">H=High</td><td>VH=Very High</td></tr>
<tr><td colspan="3">WIND=Direction/Speed(mph) 10 min average</td><td colspan="2">E=Extreme</td><td></td></tr>
<tr><td colspan="2">FFMC=Fine Fuel Moisture Code</td><td>DC=Drought Code</td><td colspan="2">ISI=Initial Spread Index</td><td>BUI=Buildup Index</td></tr>
<tr><td colspan="3">FL=Flame Length (feet)</td><td colspan="3">ROS=Rate of Spread (chains/hour or multiply by 1.1 to get feet/min</td></tr>
<tr><td colspan="6">FIRE OBSERVATION/DESCRIPTION-RANK (color coded)</td></tr>
<tr><td>Rank 1</td><td>Rank 2</td><td>Rank 3</td><td>Rank 4</td><td>Rank 5</td><td>Rank 6</td></tr>
</table>

Figure 15. Weather data input and CFFDRS outputs

Figures 13 and 14 show that fire weather and fire behavior conditions were severe most of the summer and early fall. Using data derived from the daily output of the CFFDRS and its Fire Behavior Prediction (FBP) system, we can make estimates of the Rate of Spread (ROS) and Flame Length (FL) for each day in the following chapter, the daily fire diary (Figure 15).

Two fuel models were used; C-2 (Boreal Spruce) and O-1b 70% cured (Standing Grass). The Rate of Spread is in chains per hour (a chain is sixty-six feet long), and flame length is in feet. The Rate of Spread is given for a fire running with the wind only, not a flanking or backing fires. These estimates are made from Fire Behavior Tables (See Appendix) taken from the Michigan Guide for CFFDRS Fire

Figure 16. Fire Observation and description

Behavior Prediction[83]. Also, see Figure 104 in the Appendix[84] for details on Fire Behavior Table usage.

As shown in Figure 16, Rank 1 and 2 fires can generally be attacked at the head or flanks by persons using hand tools; a hand line should hold the fire. Rank 3 fires are too intense for a direct attack on the head by persons using hand tools. Hand line cannot be relied on to hold the fire. Equipment such as tractor-plows, bulldozers, engines, and retardant aircraft can be effective. Rank 4 fires may present serious control problems-torching out, crowning, and spotting. Control efforts at the head will probably be ineffective. Rank 5 fires will exhibit crowning, spotting, and major runs are probable. Control efforts at the head of the fire are ineffective. Rank 6 fires can exhibit violent fire behavior with fireballs, fire whirls, and an organized crown fire front.

GREAT LAKES FIRE/FUELS INFORMATION

In the following chapter, daily weather data, CFFDRS codes and indexes, and fire behavior outputs are shown along with fire behavior rankings to give context to the daily struggle firefighters were engaged in to suppress the Seney fire.

A further refinement of the use of weather information was developed through the cooperation of the state natural resource agencies in Michigan, Minnesota, Wisconsin, and Mesowest[85] starting in 2008. It provides a comprehensive, web-based set of fire weather and fire danger information, integrating it with modeled fire danger, behavior, and effects information in a single information system. It provides important decision-making resources for fire managers in the Lake States. The Great Lakes Fire and Fuels Information system utilizes data from a variety of weather recording stations and provides graphic

83 Ziel, Robert, *Michigan Field Guide for CFFDRS Fire Behavior Prediction (FBP) System*, March 2015

84 National Wildfire Coordinating Group, *Fire Behavior Prediction (FBP) System*, https://www.nwcg.gov/publications/pms437/cffdrs/fire-behavior-prediction-system, accessed May 18, 2022

85 University of Utah, *Mesowest*, http://mesowest.utah.edu

	FFMC	DMC	DC	ISI	BUI	FWI
LOW	0.0-59.9	0.0-5.9	0.0-14.9	0.0-1.9	0.0-9.9	0.0-1.9
	60.0-74.9	6.0-9.9	15.0-49.9	2.0-2.9	10.0-14.9	2.0-3.9
	75.0-80.9	10.0-12.9	50.0-79.9	3.0-3.9	15.0-18.9	4.0-4.9
MOD	81.0-82.9	13.0-19.9	80.0-99.9	4.0-5.9	19.0-24.9	5.0-7.9
	83.0-84.9	20.0-24.9	100.0-199.9	6.0-6.9	25.0-29.9	8.0-9.9
	85.0-86.9	25.0-27.9	200.0-209.9	7.0-7.9	30.0-33.9	10.0-13.9
HIGH	87.0-87.9	28.0-33.9	210.0-229.9	8.0-8.9	34.0-39.9	14.0-15.9
	88.0-88.9	34.0-39.9	230.0-249.9	9.0-9.9	40.0-49.9	16.0-18.9
	89.0-89.9	40.0-41.9	250.0-273.9	10.0-10.9	50.0-53.9	19.0-20.9
VERY HIGH	90.0-90.9	42.0-49.9	274.0-299.9	11.0-13.9	54.0-61.9	21.0-23.9
	91.0-91.9	50.0-59.9	300.0-329.9	14.0-16.9	62.0-69.9	24.0-29.9
	92.0-92.4	60.0-62.9	330.0-359.9	17.0-18.9	70.0-76.9	30.0-32.9
EXTREME	92.5-93.9	63.0-69.9	360.0-399.9	19.0-21.9	77.0-79.9	33.0-34.9
	94.0-94.9	70.0-79.9	400.0-499.9	22.0-24.9	80.0-99.9	35.0-44.9
	95.0+	80.0+	500+	25.0+	100.0+	45.0+

Figure 17. Fuel Moisture and Fire Behavior Codes and Indexes

representations and a wealth of hourly weather observations.[86]

This system provides the CFFDRS database access to the complete set of weather recording stations available to Mesowest, including the National Weather Service, Remote Automated Weather Stations, and Enviroweather stations. Weather observations include temperature, relative humidity, wind speed, and accumulated rainfall. The CFF-DRS output products are calibrated to estimate fuel moisture and fire behavior for the peak fire period of the day at 1600 local standard time.

These refined thresholds, divided into fifteen categories from the original five, are calibrated to weather observations and associated fire occurrence in the northern Lake States (Figure 17).

Where possible, the upper and lower classes within each color group

---

86 *Great Lakes Fire/Fuels Information*, https://glff.mesowest.org/

suggest that the value is in transition to (or from) the adjacent color group. These tend to include narrower bands of values. It is possible to have "extreme" conditions in one code and "low" conditions in another.

SUMMARY OF CFFDRS INDEXES AND CODES

For example, spring FFMC values can quickly reach elevated levels with a few days of drying, while DMC and DC generally require weeks or months to reach levels of concern.

FFMC		DMC		DC		ISI		BUI	
87-90.9	27	28-49.9	26	210-299.9	1	8-13.9	16	34-61.9	12
91-92.4	15	50-62.9	5	300-359.9	11	14-18.9	6	62-76.9	16
92.5+	4	63+	23	360+	44	19+	2	77+	28

Figure 18. Number of days at each color level, July 30–September 22, 1976

Figure 18 is a summary of the CFFDRS codes and indexes for the period July 30, the day the fire started, through September 22, the day the fire was officially contained, a period of 56 days. It shows high, very high, and extreme categories for each code or index, using a combination of the threshold levels used by the MDNR (Figure 8) and refinement from the Great Lakes Fire and Fuels system.

All of the fuel moisture codes were in the high to extreme range for the majority of the fire. The FFMC was at a high level for 27 days and very high to extreme level for 19 days. These levels mean that there was a great possibility of ignition; spot fires over the line and from any other ignition source would have been problematic.

Levels of the DMC and the DC show that firefighters would have experienced considerable resistance to control and that mop-up would have been very difficult. At these levels live fuels become part of the fuel complex. During this period, the DC was greater than 360 for 44 days so the heavy deep fuels were available to burn until the moisture from winter snows saturated them.

The BUI was extreme for 28 days; it exceeded 120 two days during this period; the ISI was high to extreme for 24 days, together giving possibilities for the fire to spread and be very resistant to control.

CHAPTER FIVE

DIARY OF THE FIRE

A lot was going on in the country and around the world in 1976. Considerable excitement was generated about the United States Bicentennial celebrations; Michigan had a red, white, and blue license plate with stars and stripes commemorating it.

Other memorable events included the Concorde entering service, cutting transatlantic flying times to 3 1/2 hours. A year after Microsoft was formed, Steve Jobs and Steve Wozniak started Apple. Nadia Comaneci scored the first-ever perfect score in gymnastics at the Olympics. NASA unveiled the first space shuttle, the $2.00 bill was issued, and we had the first outbreak of Legionnaires Disease. In November, Jimmy Carter defeated incumbent Gerald Ford becoming the 39th president of the United States. The Dow Jones Industrial Average closed the year at 1004, the average cost of a new house was $43,400.00, gas cost fifty-nine cents a gallon, and paychecks averaged $16,000 a year.[87]

No one would have thought the forest fire season of 1976 in Michigan and the eastern Upper Peninsula would be remembered for many years to come. When the first snow covered the Seney swamp in late 1976, calls started coming into the MDNR office in Marquette asking for a speaker to come and tell the story of the Seney fire. A slide program was created based on my personal notes and maps. The following fire diary is based on that information; the fire map (Figures 19-20) is a much-improved version of the sketch maps I made in 1976.[88]

87 *The People History, What Happened in 1976 Important News and Events, Key Technology and Popular Culture*, http://www.thepeoplehistory.com/1976.html

88 Luoto, Cory, MDNR, Naubinway, MI, 2020

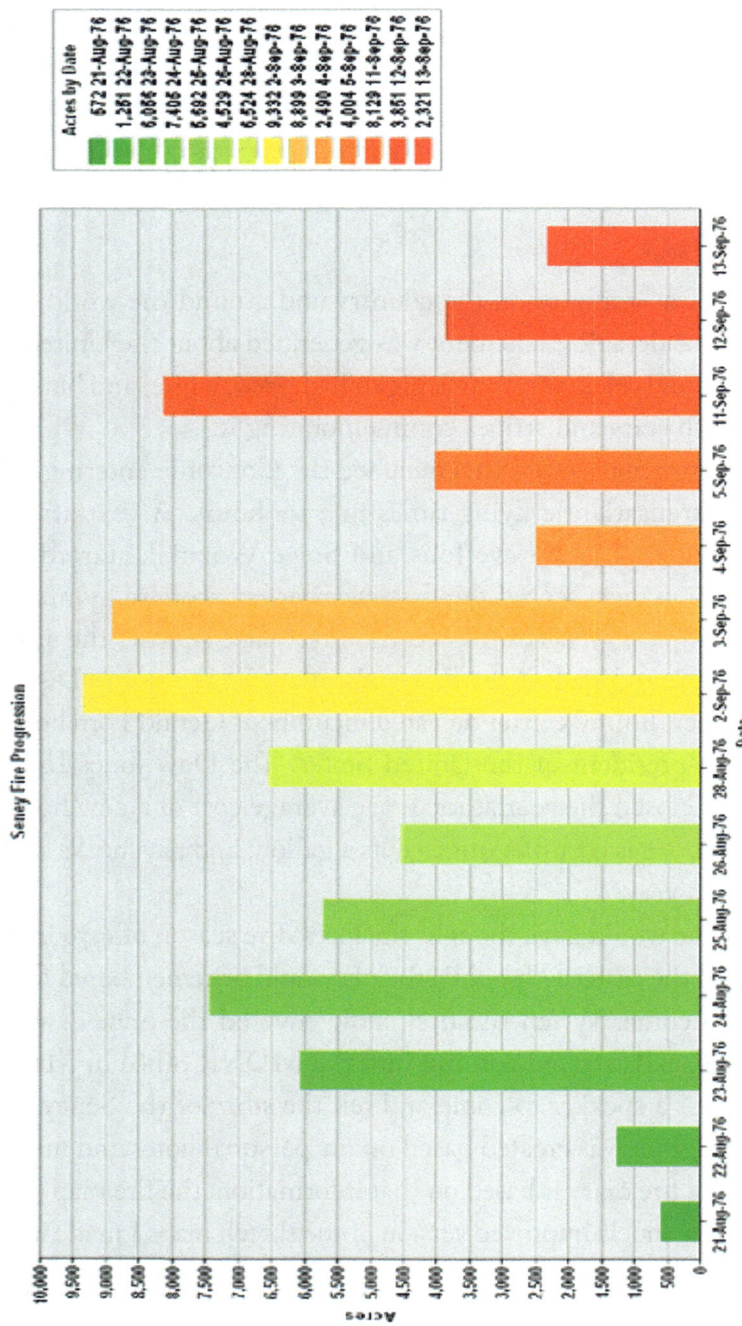

Figure 19. Approximate acreage loss by date

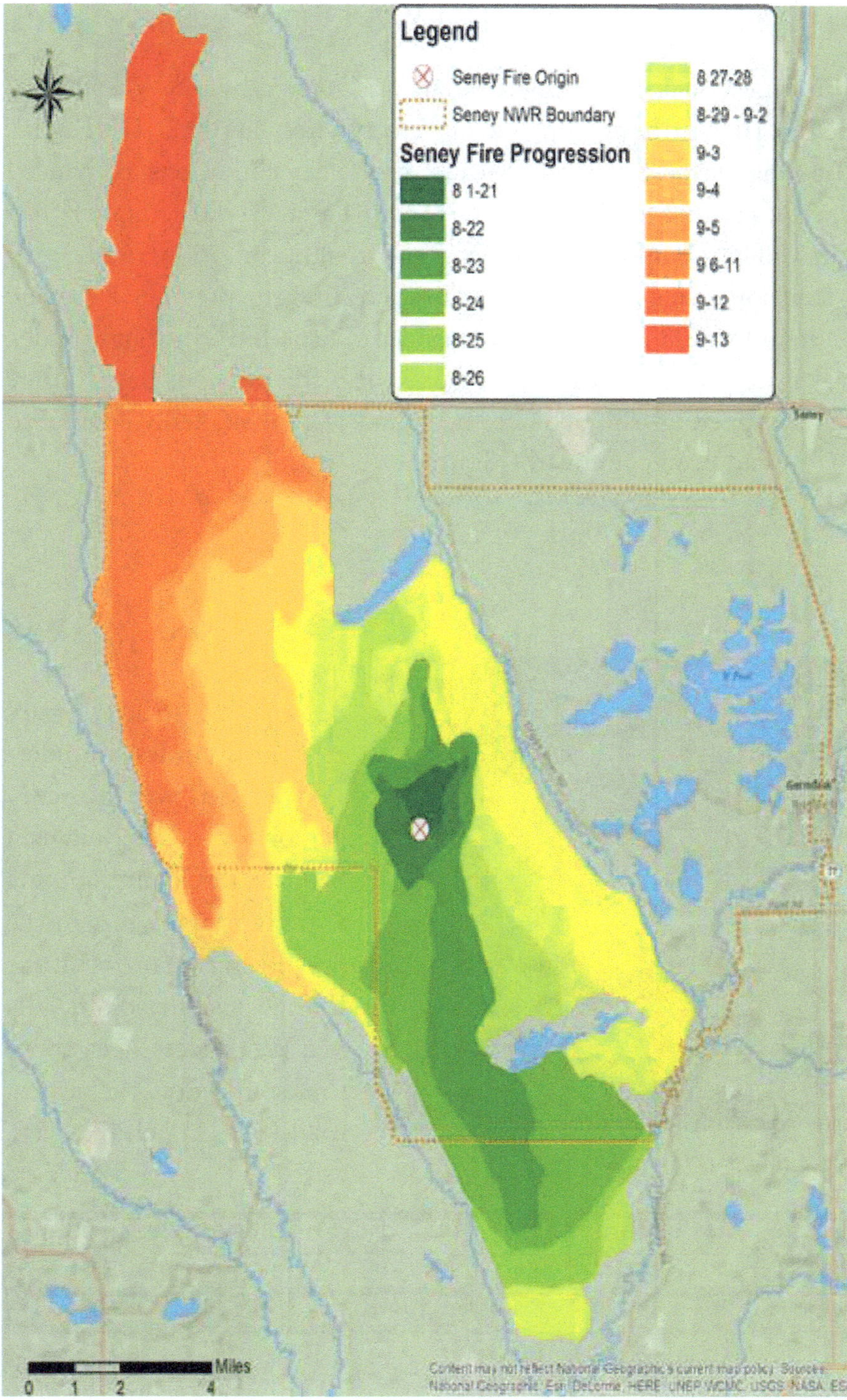

Figure 20. Seney Fire Progression Map

JULY 7

The Pine Creek "Prescribed Burn" was started by F&WS staff on Refuge lands. They planned to burn forty acres, but the MDNR staff, knowing the area well and because of the dry conditions, convinced the Refuge staff to burn only one acre first as a test burn. State Forest Fire Officers pointed out that existing drought conditions, while mitigated somewhat by 0.6 inches of rain during the previous day, were high and would cause mop-up problems. The KBDI drought index was 173.3, reduced by the rain from 202.6 the previous day. Refuge staff mopped up the one-acre burn and were mistakenly convinced that it had been contained (Figure 22).

JULY 30

Figure 21. Lightning strike

A lightning strike (Figure 21) from a rainless thunderstorm passed over the Seney National Wildlife Refuge in Schoolcraft County, igniting a fire in marsh grass in the north-central portion of the Refuge designated officially as a Wilderness Area about three miles from the nearest road (See Appendix, page 211).

TEMP	RH	WIND	FFMC	DC	ISI	BUI	KBDI	R
79	54	SE 7	74.4	298.3	1.3	46.7	207.4	H

Spruce	ROS=1	FL=1	Grass	ROS=1	FL=1

AUGUST 1

A Michigan Department of Natural Resources (MDNR) aerial detection aircraft spotted the fire shortly after noon. The pilot radioed its location to the MDNR Seney Field Office; they, in turn, relayed this to the Refuge Manager, John "Jack" Frye (Figure 23). At 1:00 p.m., from the Driggs River Road, east of the fire, he attempted to walk into the location of the fire provided by the MDNR aircraft but was not able to detect any sign of fire. About a mile and a half into the timber, he ran into a dense spruce bog thicket with high water that made further travel from this impossible.

Figure 22. Drought dried the muck and peat

Figure 23. Seney NWR Manager Jack Frye

Manager Frye returned to his vehicle and traveled to the C-3 Pool area. From there, he attempted to walk into the fire from a different direction but again was turned back by knee-deep water and dense spruce thickets; he returned to his truck around 5:00 p.m. These two attempts to locate the fire took all Sunday afternoon. Manager Frye still had not observed any smoke from the reported fire. Manager Frye's last contact with the MDNR aircraft indicated the fire was still about 1/4 acre in size and not spreading. Manager Frye attempted, unsuccessfully, on at least three occasions to walk into the fire (Figure 24).

Figure 24. Refuge staff attempted to reach the fire

Requests by the MDNR to enter the area via motorized equipment were denied due to its status as a wilderness. MDNR and Refuge air surveillance of the fire was conducted during the next ten days. District Fire Supervisor Mike Paluda, Newberry, was very concerned about the fire getting out of the Refuge and spreading to the forest. He directed Area Fire Supervisor George Young from Manistique to mark the boundary between the State and Refuge with toilet paper hanging in the trees. This enabled the pilots to easily see, every day, how close the fire was getting to the boundary. MDNR Forest Fire Division staff continued to recommend immediate control, but Manager Frye did not feel an emergency existed that warranted equipment access.

TEMP	RH	WIND	FFMC	DC	ISI	BUI	KBDI	R
70	34	NW 12	90	313.2	11.3	55.6	219.7	VH

Spruce	ROS=42	FL=18+	Grass	ROS=43	FL=8

Figure 25. Uncontrolled fire near Walsh Ditch–early August *(see page 211)*

AUGUST 2

MDNR's contract pilot, Vern Bernard, reported that the fire had spread to ten acres or less. The fire was still 1.5 miles north of State land. From the information available to Manager Frye, he believed the fire would burn itself out. That, however, would soon change. Mike Paluda, District Fire Supervisor, Newberry, MDNR, flew the area in the late afternoon and estimated the fire at 30-40 acres (Figure 25).

TEMP	RH	WIND	FFMC	DC	ISI	BUI	KBDI	R
80	30	NW 9	92	321	11.8	60.7	227	VH
Spruce	ROS=52	FL=18+	Grass	ROS=48	FL=8			

AUGUST 3

Refuge Manager Frye and Refuge Mechanic Zellar attempted again to walk into the fire, this time from the south. Although they were able to get within one mile of the fire, travel again became impossible, and both individuals returned to their vehicle exhausted.

A Fish and Wildlife Service (FWS) aircraft, on a routine administrative flight, arrived from Minneapolis and conducted a late afternoon flight that revealed the fire was approximately one mile from State lands. The burned area was estimated to be around 200 acres.

The Pine Creek fire ignited as a one-acre test for a prescribed burn by Refuge staff on July 9, 1976, and was now five acres in size. Fortunately, with expert advice from the MDNR beforehand, knowing the area and the drought conditions, the burn was originally kept small. On July 28, MDNR staff recommended that the Refuge extinguish the burn, but no action was taken for two days.

TEMP	RH	WIND	FFMC	DC	ISI	BUI	KBDI	R
82	26	SW 8	93.2	329	12.8	66.1	229.1	VH
Spruce	ROS=57	FL=18+	Grass	ROS=54	FL=8			

AUGUST 4

The Refuge staff continued to monitor the main fire with the FWS aircraft. There seemed to be no change in the size or direction of the fire. MDNR continued to monitor the main fire each day as part of their regular aerial patrol, as did the F&WS aircraft.

TEMP	RH	WIND	FFMC	DC	ISI	BUI	KBDI	R
78	43	SW 13	91.7	336.6	15.5	69.9	235.4	VH
Spruce	ROS=74	FL=18+	Grass	ROS=71	FL=10			

AUGUST 5

Figure 26. Pine Creek fire jumps the line

The Pine Creek fire jumped a ditch and started to expand rapidly (Figure 26). Manager Frye committed his entire staff of six employees to the fire. It was his feeling that this fire posed a threat and lent itself to ground equipment control. The whole Refuge crew and several State personnel and their equipment were continuously committed to this fire from August 5-17, aggressively suppressing the escaped fire until it was controlled.

TEMP	RH	WIND	FFMC	DC	ISI	BUI	KBDI	R
67	67	W 9	62.1	325.2	1	48.6	235.7	L
Spruce	ROS=1	FL=1	Grass	ROS=1	FL=1			

AUGUST 9

Bob Compeau, Chief of the Forest Fire Division, MDNR, contacted Bob McDonald, State and Private Forestry, US Forest Service, Upper Darby, Pennsylvania, expressing concern about the fire. The MDNR sent personnel on all-terrain vehicles (Figure 27) to mark the corner of State-Federal property and to gather intelligence on the area. Area Fire Supervisor George Young reported to Paluda that the marsh and fuels were very dry, and a real potential fire problem existed unless extinguished. MDNR Regional Fire Supervisor Albert "Bud" Livingston flew the fire and estimated its size at 1500 acres. Manager Frye informed District Fire Supervisor Paluda that they would maintain surveillance of the fire and take appropriate action, when necessary, to control it.

Figure 27. Bombardier tracked vehicle

TEMP	RH	WIND	FFMC	DC	ISI	BUI	KBDI	R
83	38	S 7	91.7	355.5	9.6	67.9	252.4	H

Spruce	ROS-41	FL-18+	Grass	ROS=38	FL=6

AUGUST 10

The fire size was estimated at 1200 acres in size by Bob Drieslein, Assistant Refuge Manager. As a result of telephone contacts made on August 9, Norrell Wallace of the Twin Cities Regional Office,

F&WS, called the MDNR State fire supervisor, Bob Compeau. Wallace was not fully aware of the Walsh Ditch Fire size, thinking both fires (including Pine Creek) were about 200 acres each. The F&WS recognized the large fire (Figure 28) potential could exceed the guidelines set down in their wilderness management policy.

Figure 28. Uncontrolled fire in the Seney swamp, early August

TEMP	RH	WIND	FFMC	DC	ISI	BUI	KBDI	R
80	45	S 14	91.1	363.3	15.6	71.9	260.9	VH

Spruce	ROS=74	FL=18+	Grass	ROS=71	FL=10

AUGUST 11

Figure 29. The fire continues to spread

The main fire (Figure 29) was still approximately one-half mile north of State land, with the greatest spread occurring on the north side. That afternoon Manager Frye called the MDNR, asking what action they should take to protect State land. He was advised that he would need at least two large bulldozers, as there was a danger of getting stuck, and one would probably be required to pull the other out. The Pine Creek fire was contained at 200 acres.

TEMP	RH	WIND	FFMC	DC	ISI	BUI	KBDI	R
82	59	SW 11	89.1	371.3	9.2	75.1	269.3	VH

Spruce	ROS=36	FL=18+	Grass	ROS=32	FL=6

AUGUST 12

Figure 30. Contract bulldozers

MDNR Area Fire Supervisor George Young had contract bulldozers (Figure 30) start toward the fire site. That evening, District Fire Supervisor Paluda advises Manager Frye that, to sustain the control effort, the Refuge will have to provide support services and supplies such as reserve gasoline and medical aid. The fire was estimated at 1500 acres.

TEMP	RH	WIND	FFMC	DC	ISI	BUI	KBDI	R
80	61	S 5	67.8	359.7	0.9	53.1	270.6	L

Spruce	ROS=1	FL=1	Grass	ROS=1	FL=1

AUGUST 13

After checking with superiors in Minneapolis and Washington, Frye told Paluda that the State control effort was to be called off. Young and contract bulldozers turn around without having reached the fire. One bulldozer became mired in the swamp and was not extricated until three days later. Paluda advised Frye that the MDNR could not further assist with the fire inside the Refuge under the circumstances. In the evening, FWS officials in Minneapolis called the Boise Inter-Agency Fire Center (BIFC) for advice and assistance.

TEMP	RH	WIND	FFMC	DC	ISI	BUI	KBDI	R
66	37	NW 6	83.6	366.1	2.8	56.8	277.2	L

Spruce	ROS=8	FL=8	Grass	ROS=6	FL=4

AUGUST 15

The fire increased in size to the northeast and was stable in the south, and was approximately one-half mile from State land. Refuge, MDNR personnel, and bulldozer operators continued with attempts to move equipment out of the area. MDNR was still monitoring the fire from the air as part of regular surveillance and fire detection flights. Assistance was requested from the BIFC Fire Cache by the F&WS.

TEMP	RH	WIND	FFMC	DC	ISI	BUI	KBDI	R
67	34	N 8	89.9	379	8	64	284.3	H

Spruce	ROS=30	FL=12	Grass	ROS=27	FL=6

AUGUST 16

Greg Jensen of BIFC arrived on the scene and flew over the fire; he estimated it at 1800 acres in size. Jensen returned to the Refuge, contacted BIFC, and arranged for support. There were still doubts concerning funding. BIFC referred Jensen to the US Forest Service, Region 9, Milwaukee, where he found some firefighters and equipment available. Their only helicopter under contract was committed to initial attack on the Superior National Forest fires in Minnesota.

TEMP	RH	WIND	FFMC	DC	ISI	BUI	KBDI	R
75	35	W 7	90.7	386.3	8.3	68.4	286.5	H

Spruce	ROS=31	FL=18+	Grass	ROS=43	FL=8

AUGUST 18

John Russell took over as the Federal Fire Boss. He flew to the fire area early in the morning.; found that the nearest access for equipment was still five miles from the fire. Paluda of the MDNR advised that a Bombardier tracked vehicle could continue to be used until needed by the State. Effects of the deepening drought, low humidity, and strong winds increased control difficulties.

TEMP	RH	WIND	FFMC	DC	ISI	BUI	KBDI	R
76	56	S 13	89.2	401.5	10.9	76.3	296.1	VH

Spruce	ROS=47	FL=18+	Grass	ROS=43	FL=8

AUGUST 19

Two contract bulldozers and two BIFC fire crews, with 40 firefighters, began constructing a fire line on the southwest flank of the fire to prevent its spread onto the Manistique River State Forest to the south. A third bulldozer was put into action on August 21. Although BIFC responded immediately by sending Greg Jensen to inspect the fire and recommend a course of action, no active suppression efforts were resumed until August 19. Thus, six days were lost, and by this time, the fire exceeded 1800 acres in size. The two rental bulldozers began to construct a road on the east side of Walsh Ditch (see page 211) to a temporary fire camp.

Figure 31. Bulldozer clearing fireline

TEMP	RH	WIND	FFMC	DC	ISI	BUI	KBDI	R
85	54	S 16	89.2	409.8	14	80	300	VH

Spruce	ROS-65	FL-18+	Grass	ROS=59	FL=12

AUGUST 20

A 20-person Hiawatha National Forest hand crew arrived on fire. The crew, along with a bulldozer (Figure 31), began construction of a fire line, holding the line with hose lays and water on the west side of Walsh Ditch to stop the fire from spreading westward toward State land.

TEMP	RH	WIND	FFMC	DC	ISI	BUI	KBDI	R
85	42	S 14	90.3	418.1	13.8	84.5	308.7	VH

Spruce	ROS-64	FL-18+	Grass	ROS=59	FL=12

AUGUST 21

Two hand crews arrived from BIFC. These two hand crews and two bulldozers were added to the fire line construction on the southwest flank of the fire. An additional bulldozer was rented.

TEMP	RH	WIND	FFMC	DC	ISI	BUI	KBDI	R
81	42	S 12	90.3	426	11.8	88.7	319.3	VH

Spruce	ROS= 55	FL=18+	Grass	ROS=48	FL=8

AUGUST 22

The fire size was approximately 2400 acres in the morning. Low relative humidity continued drying fuels in the swamp, coupled with windy conditions, which made it conducive to burning.

TEMP	RH	WIND	FFMC	DC	ISI	BUI	KBDI	R
82	28	N 8	92.6	434	11.8	93.8	327.2	VH

Spruce	ROS=55	FL=18+	Grass	ROS=48	FL=8

AUGUST 23

At 9:00 a.m., MDNR aerial surveillance found that the fire, whipped by strong northerly winds with warm temperatures and low humidity, ran nine miles southeast during the past 24 hours. It jumped the partially completed fire lines and spilled over the Refuge boundary, running two to three miles onto State land. Extreme burning conditions continued until approximately 2:00 a.m. on August 24.

The fire's huge run burned an estimated six sections, nearly 4000 acres, of State land. Orders for equipment, firefighters, equipment operators, and support staff were made throughout the Upper Peninsula and Northern Lower Michigan.

Mike Paluda, District Fire Supervisor for the eastern Upper Peninsula, was put in charge as Fire Boss for the State and took immediate action (Figure 32). Fire suppression continued throughout the remainder of the week, with equipment and human resources coming from every Division in the Department and from all parts of the State.

Figure 32. Mike Paluda, MDNR Fire Boss

Figure 33. Convection column

As it grew, the fire (Figures 33-34) was contained by fire lines that were constructed on the south and east sides by bulldozers and MDNR tractor-plows. Although it was a continuous fight to hold them, the fire lines, for the most part, held. Ten fire crews, 20 firefighters per crew, and a Type I overhead team were requested from the Boise, Idaho Inter-Agency Fire Center (BIFC) and were working on the fire by the end of the week. State and Federal fire camps were established at the Cookson Bridge crossing of the Manistique River.

The BIFC "spike" camp inside the Refuge was evacuated. BIFC headquarters was notified of the fire breakout and dispatched a Department of Interior fire overhead team with Bob Webber as the Fire Boss. They were one of the first of six national Type I overhead teams created that year to manage large fires.

By the end of the day, there were 35 pieces of large MDNR equipment working on the fire, including tractor-plows, bulldozers, water tankers, pumpers, and aircraft, being staffed by 100 MDNR personnel.

Figure 34. Fire runs nine miles and onto State land

Figure 35. Equipment was brought from all over the State

Figure 36. Using a tractor-plow to construct fireline

(Figure 35)

The terrain, soils, and vegetation caused numerous problems. The organic soils, a combination of peat and muck, sometimes several feet deep, were burning on and beneath the surface. There were pools of water, streams, and small rivers with areas of marsh grass interspersed with ridges and islands of pine, hemlock, spruce, ash, and birch. Most of the land did not lend itself well to travel, even by four-wheel-drive vehicles. Access, in many cases, was only by high-flotation tracked vehicles.

There was a continuous struggle on the line by personnel and equipment to prevent further acreage loss. Fire line construction (Figure 36) through the wetland was difficult, at best. The crews had to dig down to mineral soil, removing deadwood and any roots that were burning like a fuse to unburned fuel on the other side of the line. Once the lines were built, they were burned out in an attempt to remove unburned fuel from the path of the oncoming fire.

The entire MDNR Marquette regional repair shop staff (Figure 39-40) was on the scene at one time, plus staff from the Gaylord shop.

Figure 37. State fire camp at Cookson Bridge

Figure 38. State and federal forces co-locate fire camps

Figure 39. MDNR mechanics Reino Lintula and John Jaykka

Figure 40. MDNR mechanic Bill Reynolds

TEMP	RH	WIND	FFMC	DC	ISI	BUI	KBDI	R
75	48	NE 5	90.8	441.3	7.1	97.1	333.8	H

Spruce	ROS-27	FL-18+	Grass	ROS=22	FL=4

AUGUST 24

Firefighters attacked large amounts of open or unsecured lines; work continued with equipment to build fire lines and contain the escape.

Figure 41. Bob Webber, USDI Fire Boss

Figure 42. Supplies being received at Marquette, August 25

The US Department of Interior (USDI) Overhead Team, with Bob Webber as Fire Boss (Figure 41), arrives and takes charge of the Federal control effort. The fire was now more than 20,000 acres and growing. MDNR crews were concentrating on building containment lines (Figure 36) on the south end of the fire.

The Federal fire camp was established adjacent to the State camp at Cookson Bridge (Figure 37-38). BIFC supported both Federal and State crews, a massive logistical operation; incoming supplies were measured by the ton (Figures 42-44).

Figure 43. Fire pumps

Figure 44. Camp Boss and supply

TEMP	RH	WIND	FFMC	DC	ISI	BUI	KBDI	R
80	38	S 8	90.8	449.1	9.2	101.3	339.6	VH

Spruce	ROS=38	FL=18+	Grass	ROS=32	FL=8

AUGUST 25

Figure 45. NG trucks hauled personnel and supplies

Figure 46. 66 NG trucks put on 45,000 miles

Figure 47. NG worked 24 hours a day feeding firefighters

After 72 hours of non-stop work, MDNR firefighters completed containment lines around the south end of the fire. Governor Milliken declared an emergency allowing the Michigan National Guard to move in to feed and transport firefighters. Most wondered how the fire operations had gotten along without them.

The 107th Engineer Battalion, with armories across the Upper Peninsula, assumed the mission of forming a guard task force. They provided critical motor transport for firefighters into the fire area, fed the entire crew, hasty combat construction of roads and bridges in the fire zone, and controlled their helicopter assets for aerial inspection and monitoring.

Sixty 2½ ton National Guard trucks (Figures 45-46) from the 107th Engineer Battalion and 1437th Engineer Company eventually put on over 45,000 miles hauling personnel and equipment.

Feeding everyone on the fire lasted almost two months (Figures 47-51). Guard cooks worked 24 hours a day to provide over 65,000 meals at the cost of $152,000. Once the guard started cooking, the complaints about food virtually disappeared. Firefighters could go through the chow line as often as they liked. Bag lunches were issued each morning to be taken and

Figure 48. Food bill is over $7,000 a day

Figure 49. Firefighters eat standing up

Figure 50. The daily shopping list was huge

Figure 51. Food was available day or night

eaten on the fire line.

It was an immense operation, as evidenced by one breakfast that fed 1050 people over 7000 pancakes. At one point, over $7,000 a day was being spent on food for the firefighters, a bill that exceeded the combined weekly shopping lists of all the restaurants in Marquette.

Preparing the meals involved three eight-hour shifts that had to plan for 800 meals a shift. Shopping lists were huge. Breakfast: Two cases of tomato juice, four cases of dry cereal, 120 dozen eggs, two cases of pancake mix, two cases of maple syrup, 100 pounds of bacon, and two cases of potatoes and hash browns. Lunch: 200 pounds of lunch meat, 1,400 candy bars, 800 packs of gum, 800 individual bags of potato chips, 400 pounds of apples, oranges, and bananas, 200 cases of soda pop, and six jars of tea. Dinner: 200 pounds of turkey, a case of gravy mix, two cases of dry-mix potatoes, three cases of cranberry sauce, five cases of asparagus, three cases of lettuce, a case of salad dressing, three cases of sweet cherries, and three cases of assorted cookies. Plus, 245 loaves of bread, 25 pounds of butter, 30 pounds of coffee, and 2,400 pints of milk.

If anybody went hungry, it was their fault. There was always something to eat and plenty of coffee, day or night.

Figure 52. Guard equipment and personnel

When three bridges in the fire area became impassable due to the heavy traffic, the 107th Engineer Battalion provided both the personnel and heavy equipment needed to repair them quickly (Figures 52-53). Two of the bridges were redecked with heavy timbers, and the third was replaced with a large culvert.

Figure 53. The 107th repaired three bridges

Three Guard helicopters (Figure 54), backed by helicopters from the Regular Army and the Department of Interior, provided quick reaction capability for the fire crews. This enabled the firefighters to reach "hot spots" that could not be reached by truck.

At the fire's height, 147 members of the 107th were on duty, and by the fire's end, 550 personnel from 13 units had seen fire duty; the assignment lasted almost eight weeks, assisting on the fire until October 15. The 107th Engineer Battalion's support of the Seney Fire personnel was critical to the suppression operations' success.

Figure 54. Three NG helicopters were on the fire

There was a massive build-up of Federal forces to an eventual strength of 750 firefighters plus support personnel underway. Aerial water-bombing capability, both fixed-wing and helicopters, were also called in.

Figure 55. Schoolcraft County Road Commission on scene

Additional logistical problems were solved by cooperation with other agencies and the private business sector. The Schoolcraft County Road Commission (Figure 55) helped with road and bridge repair and culvert replacement. Fuel supplies were purchased from local oil companies. Clean drinking water was hauled in milk trucks (Figure 56) at 1,000 gallons per load. Charter buses (Figure 57) were used to haul firefighters to and from the fire line from the fire camp.

Figure 56. Milk trucks hauled drinking water

Figure 57. Charter buses hauled firefighters

TEMP	RH	WIND	FFMC	DC	ISI	BUI	KBDI	R
84	35	SE 14	91.6	457.3	16.6	106	345	VH

Spruce	ROS=86	FL=18-	Grass	ROS=76	FL=12

AUGUST 26

The Michigan State Police moved a command trailer into the fire camp to coordinate emergencies, communications, and highway closures (Figure 58).

Work continued constructing fire lines around a 56-mile perimeter, with MDNR supplying equipment and operators and BIFC providing crew personnel (Figures 59-61)

The total number of personnel in the fire camp rose above 1,000. Firefighting crews and equipment were drawn in from 29 states, including Alaska. Burnout operations began, back from the lines toward the fire.

Radio communication problems on the fire were unique. The low-band radio system operated by the MDNR and Refuge was often

Figure 58. Michigan State Police coordinated highway closures.

Figure 59. Bulldozer clearing fuel off of the fireline.

Figure 60. MDNR tractor-plow making a plow line.

Figure 61. Two bulldozers clearing the fireline.

compromised by heavy brush and vegetation. MDNR radio technicians worked tirelessly to establish and maintain repeaters and do necessary repairs to provide essential coverage. The entire system had only four low-band frequencies. The lack of tactical channels required exceptional radio discipline for everyone using the system. This impacted MDNR communications in northern lower Michigan as well. The Federal forces had their own high-band radio system and fewer problems with radio communications.

Mutual aid came from fire departments across the Upper Peninsula (Figure 62).

TEMP	RH	WIND	FFMC	DC	ISI	BUI	KBDI	R
80	51	SE 9	90.3	465.1	9.2	109.3	350.6	VH

Spruce	ROS=39	FL=18+	Grass	ROS=32	FL=8

Figure 62. Mutual aid came from across the UP. Cooperators included: Sault Ste. Marie (2 trucks), Negaunee, Marquette, Munising, Newberry, Menominee (2 trucks), Iron River (2 trucks), and Stephenson

AUGUST 28

The fire grew to 22,000 acres in size. The fire jumped over the MDNR line on the southeast flank but was contained with aerial water drops after burning 15 acres of State land. This was the first demonstration that the fire could "come alive" and "reburn" over previously burned marsh vegetation and organic soils that were continuously drying out because of the drought. The escape was contained with ground units and water drops (Figure 63-67). A decision was made to unify

Figure 63. Aerial water drops keeps the fire from spreading

Figure 64. Firefighter mopping up the escaped fire

command into a joint State-Federal structure with co-Fire Bosses (Paluda and Webber) and integrated staff and line supervisors, equipment, and firefighters to be fully interchangeable.

TEMP	RH	WIND	FFMC	DC	ISI	BUI	KBDI	R
73	30	S 11	87.8	480.4	7.6	114	366.2	H
Spruce	ROS=34	FL=18+	Grass	ROS=27	FL=6			

Figure 65. Technicians mounting an infrared scanner

AUGUST 29

Infra-red mapping (Figure 65) indicated that the fire was 30,850 acres in size. Suppression actions were concentrated on fire line completion and strengthening in preparation for burning out to the control lines. The southern point of the fire was again contained.

Figure 66. Handcrews continued to improve fireline

Figure 67. Mop-up was difficult due to the extreme drought

TEMP	RH	WIND	FFMC	DC	ISI	BUI	KBDI	R
52	41	W 18	87.8	485.4	13.5	115.8	372.2	VH

Spruce	ROS=69	FL=18+	Grass	ROS=59	FL=12

AUGUST 30

The fire crept in a northwesterly direction; its size was estimated at 38,500 acres. Containment of the fire was planned for the afternoon. Winds again caused spotting problems; one spot is over highway M-28.

TEMP	RH	WIND	FFMC	DC	ISI	BUI	KBDI	R
66	41	SE 9	88.6	491.8	7.2	118.6	374.9	H

Spruce	ROS=28	FL=18+	Grass	ROS=22	FL=4

AUGUST 31

The fire size was now 40,000 acres. Fire lines were well established and burned out except for the north and west flanks. Consultation with specialists and resource managers resulted in a decision to burn out the north and west sides of the Refuge. The area to be burned was approximately 35,000 acres when weather and fuel conditions were optimum. The Pine Creek fire was declared out. A small stream was diverted to assist in suppressing the fire. Less than one inch of rain fell in August.

TEMP	RH	WIND	FFMC	DC	ISI	BUI	KBDI	R
64	32	NW 7	90	498	7.5	121.6	376.6	H

Spruce	ROS=34	FL=18+	Grass	ROS=27	FL=4

SEPTEMBER 1

The fire grew to 41,000 acres in size. Fire crews continued to mop up on the south and east sides, improving lines and preparing for the burnout when weather conditions were right.

Gov. William G. Milliken ordered a ban against smoking, campfires, and burning of brush, debris, and dead vegetation on private or public lands in the northern two-thirds of Michigan.

TEMP	RH	WIND	FFMC	DC	ISI	BUI	KBDI	R
53	48	W 12	74.2	503.1	1.9	106.1	378.5	L

Spruce	ROS=5	FL=8	Grass	ROS=3	FL=2

SEPTEMBER 2

Some minor scale burning was completed along control lines, limited by high humidity. The fire size was 41,965 acres.

TEMP	RH	WIND	FFMC	DC	ISI	BUI	KBDI	R
62	42	SE 10	84.8	507.8	4.6	108.3	379.8	H

Spruce	ROS=18	FL=18	Grass	ROS=13	FL=4

SEPTEMBER 3

The fire size is an estimated 49,000 acres. The fire's southwest corner is burned out; it runs quite rapidly towards State highway M-28, causing some concern that it may jump the hardtop road – it stops, however.

TEMP	RH	WIND	FFMC	DC	ISI	BUI	KBDI	R
70	41	SE 20	88.6	513.3	17.7	111.1	381	VH

Spruce	ROS=89	FL=18+	Grass	ROS=82	FL=10

SEPTEMBER 4

Massive burnout operations were started to consume combustible fuels remaining inside the containment lines. Mutual aid was rendered by fourteen fire departments from across the Upper Peninsula; personnel and equipment were on alert during the burnout on M-28, watching for spot fires. Ten fire department and MDNR pumper trucks patrolled M-28 vigilantly (Figure 62). The Soo Line Railroad provided an

Figure 68. Soo Line Railroad speeder patrol

18-man crew and a speeder patrol (Figure 68) with a pumper unit to protect the safety of the timber bridges on their rail line immediately

Figure 69. Over 240 firefighters were spread out on M-28

south and parallel to M-28. Twelve 20-person crews (Figure 69) were interspersed along the seven-mile-long front adjacent to State highway

Figure 70. Bulldozers and tractor-plows standby by on M-28

M-28, along with several tractor-plows and bulldozers (Figure 70-73). Two air tankers and several helicopters provided aerial support.

Considerable effort was expended removing the standing timber, for seven miles, between the railroad tracks and State highway M-28; all the old junk railroad ties on the fire side of the railroad right-of-way were moved, by hand, to the side away from the fire.

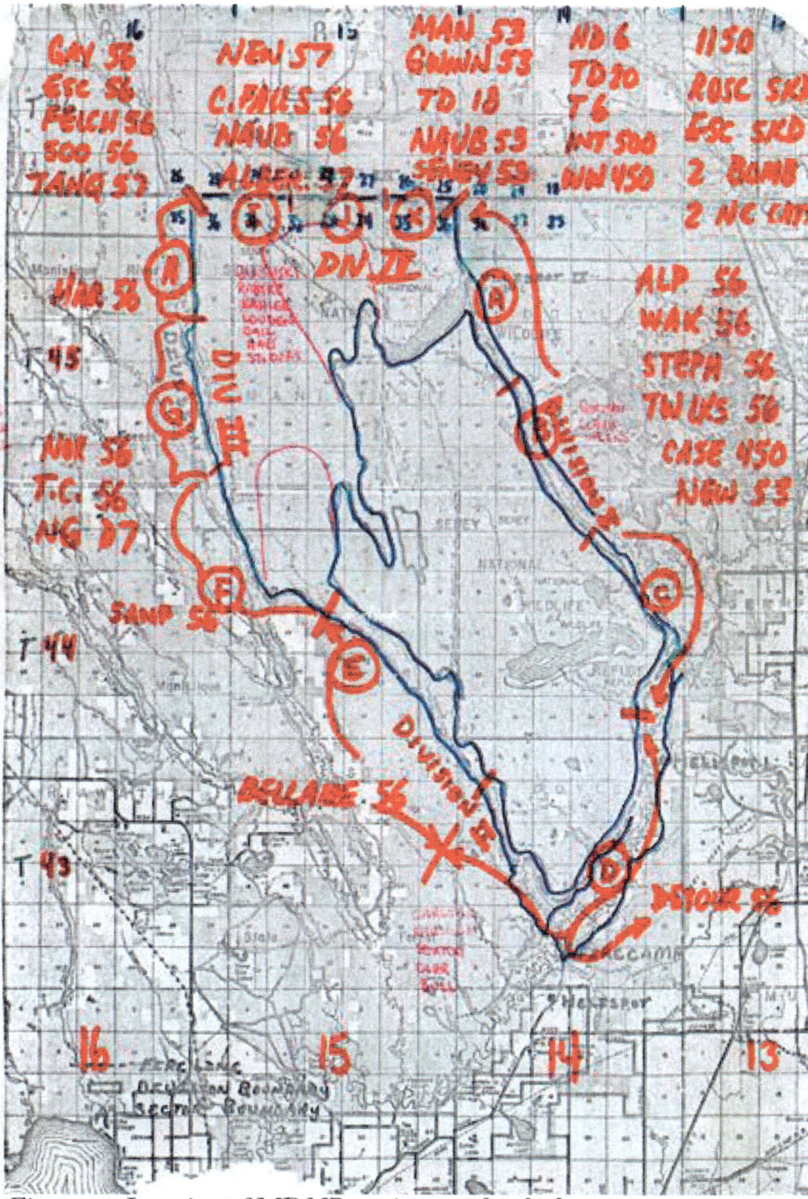

Figure 71. Location of MDNR equipment for the burnout

Nomenlature used for this map was home location (i.e., NEW is Newberry, ALP is Alpena, etc.) followed by a two digit number; 53 is a tractor-plow, 56 and 57 are 4x4 pumper trucks with tank and pump. Other units shown are skidders with plow and tank, bombardier tracked units and bulldozers.

TEMP	RH	WIND	FFMC	DC	ISI	BUI	KBDI	R
59	63	W 9	81.7	517.7	2.8	112.4	383.5	M

Spruce	ROS=9	FL=12	Grass	ROS=6	FL=3

SEPTEMBER 5

Figure 72. North Carolina firefighters

Figure 73. Firefighters and trucks on the fireline

The fire size was estimated at 54,000 acres. The weather held, and the massive backburn continued. Firefighters continued burnout operations along the Creighton Truck Trail on the west side and between the Soo line railroad tracks and State highway M-28 depriving the fire of combustible fuels that remained inside the earlier constructed containment lines (Figure 74).

TEMP	RH	WIND	FFMC	DC	ISI	BUI	KBDI	R
62	37	NW 8	87.4	522.4	5.6	114.9	386.7	H

Spruce	ROS=23	FL=18	Grass	ROS=18	FL=4

SEPTEMBER 6-LABOR DAY

I watched the Federal Operations Chief, Ted Rex, preparing to ignite the remainder of the interior of the fire from a helicopter. He came

Figure 74. Looking southeast after igniting the burnout. Source: Blanchard Mills, MDNR

over and asked me if I wanted to do the burnout since I would be here long after he returned home; that I should do it. After a quick training session with Ted on the system's mechanics and the pilot's safety protocols, I was ready to go. The system was two trays, installed under the helicopter, loaded with incendiary grenades, each approximately the size of a 12-ounce pop can (Figure 75), that could be dropped, one by one, as needed. The pilot had a mechanical safety release that could jettison the entire apparatus if safety concerns dictated. I had a bit of apprehension until I launched the first incendiary device. The aircraft was moving relatively slowly, about 25 miles per hour. The idea was to ignite the large, unburned areas with tall standing marsh grass or other fine fuels. It took a little practice to "lead" the target areas; I soon got within 5 to 10 feet of where I was aiming. We had to return several times to the landing zone to reload before the job was done. Weather and humidity permitted burning as planned. A significant effort was made to complete the planned burnout using both ground and aerial ignition. The fire size increased to 64,000 acres due to burnout operations (Figure 74).

Figure 75. Incendiary grenade

After the fire burned through the swamp grass and fine fuels, I noticed many linear features that were not visible until after the burnout. They extended in an east-west direction along the entire north end of the fire, just south of highway M-28; they resembled a braided stream channel. I discussed this with several people, and the wildlife biologists were

Figure 76. Braided caribou migration trails in the NW Territories

confident that they were remnant caribou migration trails (Figure 76) from prehistoric times.[89]

TEMP	RH	WIND	FFMC	DC	ISI	BUI	KBDI	R
72	42	SE 16	89	528.1	13.4	117.7	388.3	VH

Spruce	ROS=63	FL=18+	Grass	ROS=54	FL=8

SEPTEMBER 7

The fire was declared "contained," with a loss of 64,000 acres, the most extensive in Michigan since 1908. All but about 10,000 acres of the loss was inside the Seney Refuge; the rest was on State land to the south. An area of approximately 4,800 acres of mature timber, including 80 acres of virgin hemlock, in the northeast corner of the Refuge, was preserved from "burning out" at the suggestion of MDNR equipment operators who said they could save it and did. Demobilization began, with the Federal workforce reduced to 250 personnel and the State effort cut by about half. The fire camp location was shifted to Star Siding (Figure 77) on the north side of M-28, about five miles east of Shingleton.

Figure 77. Star Siding camp

TEMP	RH	WIND	FFMC	DC	ISI	BUI	KBDI	R
81	45	S 7	89.4	534.7	6.9	121	390.9	H

Spruce	ROS=28	FL=18	Grass	ROS=22	FL=8

89 Harper, Francis, *The Barren Ground Caribou of Keewatin*, Lawrence, Kan.: University of Kansas, 1955.

SEPTEMBER 8

The north and northeast sides of the fire were still quite hot, but no hotspots were found outside the fire line. The remainder of the fire was contained. Mop-up was in full progress, concentrating on the north, northeast, and northwest sides of the fire. A Barnes infra-red scanner was mounted on a helicopter (Figure 65) and used to find hot spots. More crew demobilization was begun.

TEMP	RH	WIND	FFMC	DC	ISI	BUI	KBDI	R
89	31	SE 12	92.7	542.1	16.6	125.6	396	E

Spruce	ROS=88	FL=18+	Grass	ROS=76	FL=10

SEPTEMBER 9

The fire was declared controlled. The USDI overhead team returned to Boise; Mitch Wiles took over as Federal Fire Boss. Mop up continued with 365 firefighters.

TEMP	RH	WIND	FFMC	DC	ISI	BUI	KBDI	R
60	68	W 12	72.3	546.6	1.8	108.5	403	L

Spruce	ROS=2	FL=5	Grass	ROS=3	FL=2

SEPTEMBER 10

Mop up operations progressed with 330 firefighters.

TEMP	RH	WIND	FFMC	DC	ISI	BUI	KBDI	R
65	43	W 15	85.2	551.6	7.3	110.9	409.9	H

Spruce	ROS=28	FL=18	Grass	ROS=22	FL=4

SEPTEMBER 12

Figure 78. Fire jumps highway M-28 Source: MDNR

Again fanned by high south winds and encouraged by low humidity, the smoldering fire in the Seney swamps came to life and raced over previously burned ground and formed a strong convection column. Firebrands were borne high into the air and blown north across M-28 (Figure 78), igniting two new fires. One was quickly contained at 300 acres in size; the other, started one-half mile north of the road and made a seven-mile overnight run northward on a two-mile front, engulfing more than 5,000 acres. Relative humidity readings I took were in the low 30 percent range at midnight, considered dry for a mid-day reading, made conditions very conducive to burning. The new "North" fire burned well into the early morning of the 13th.

TEMP	RH	WIND	FFMC	DC	ISI	BUI	KBDI	R
84	27	SW 12	92.9	564.5	17.1	118.8	417.3	E

Spruce	ROS=87	FL=18+	Grass	ROS=76	FL=10

SEPTEMBER 13

The North Fire expands to the east, increasing in size to 8,400 acres. The fire, again, is declared "major" by BIFC, and an Area Command team was formed (in ICS, this is called a General Headquarters or GHQ) with co-commanders Robert Sellers (BIFC) and Bob Compeau, the MDNR State fire supervisor. The fire effort is divided

L to R: Leon Guzinski-Division Supervisor, Robert Johnson-Fire Officer, John Levanen-Fire Officer, Greg Lusk-Assistant Fire Boss, John Robertson-Resource Coordinator

Jack Kintz
Division Supervisor

L to R: Robert "Bud" Kingren-Division Supervisor, Dan Farnsworh-Resources

John Johnston
Fire Officer

L to R: Fire Bosses Mike Paluda and Bob Webber

Mike Paluda, Fire Boss, briefing firefighters before they start their day

Figure 79. Fire planning was the key to success

L to R: Greg Lusk-Assistant Fire Boss, John Snyder-Services, Carol Schmidt-Dispatcher

Matthew "Red" Skovera Timekeeping

John Robertson-Resource Coordinator, Jim MacGregor-Plans, Greg Lusk-Assistant Fire Boss

Howard Rice-Line Boss

Jim Noasconi-Division Supervisor

George Young-Division Supervisor

Figure 80. Personnel from every division of the MDNR assisted into two zones, with State highway M-28 as the dividing line. Mike Paluda is the Fire Boss in charge of the State fire effort for the new North Zone, and Mitchell Wiles, relief for Webber, is named Fire Boss for the South Zone's Federal forces. Area Command provided coordination between the two zones and helped relieve governmental pressure from the Fire Bosses. Planning was the key to success (Figures 79-80). Information Officers handled a call every two minutes; their office is staffed twenty-four hours a day.

TEMP	RH	WIND	FFMC	DC	ISI	BUI	KBDI	R
78	43	SE 19	91.5	570.8	24.7	122.1	422	VH

Spruce	ROS=135	FL=18+	Grass	ROS=119	FL=12

SEPTEMBER 14

The remobilization of Federal and State forces brought personnel numbers back to over 1,000, they were housed in the Star Siding fire camp located north of State highway M-28.

Aerial water-dropping capability is recalled and beefed up with two fixed-wing PBY5As (Figure 81), DC-4s, and helicopters with water buckets. With the air force reestablished, outbreaks were quickly doused with retardant or water. An Air Attack Boss directed aerial activities, and was responsible for the "air force." Large helicopters (Figure 82) were assigned, including a CH-47 Chinook and a CH-54 Skycrane from Fort Sill, Oklahoma, that were capable of drafting from water sources and delivering thousands of gallons of water per hour to the fire.

At one point, there were over twenty aircraft, including nine helicopters, assigned to the fire. Three helicopters with buckets and the PBY5As dropped thousands of gallons of water, assisting the firefighters on the ground on both the North and South Zone fires. A partial list of the type of aircraft on the fire includes the following.

Figure 81. Two PBY5A" Catalinas" water bombers dropped thousands of gallons of water and retardant.

A CH-47 helicopter came from Ft. Sill, Oklahoma. It was equipped with a bucket and could draft from water sources near the fire and hauled over 2500 gallons of water per load.

A CH-54 Skycrane, from Fort Sill, Oklahoma, capable of hauling 2500 gallons of water per load is shown here loading from the ditch alongside State highway M-28.

The National Guard helicopters hauled personnel and equipment to the fireline.

Figure 82. Large helicopters including a CH-47 Chinook and a CH-54 Skycrane were assigned to the fire

Helicopters: Bell 47G2 equipped with IR scanner; Bell 206, one private aircraft (used as an aerial command post and for aerial ignition), and three military aircraft (primarily used for hauling personnel and equipment); a Bell 212 equipped with a tank for water bombing and crew movement; CH-47 (two aircraft); and a CH-54, all equipped with buckets for water bombing (Figure 82). Fixed-wing aircraft included: Cessna 180 (observation and detection), Beaver w/floats (equipped with a tank for water bombing), Baron (used as a lead plane for air attack, recon, and detection), PBY5A (two aircraft used for retardant and water bombing), two DC-4's (retardant bombing), a Cessna 310 (lead plane) and a King Air (equipped with IR scanner).

Aerial resources on a wildfire maintain a high degree of safety in their operations. On one occasion, it came as a large surprise to have a SAC B-52 bomber on a training mission from KI Sawyer Air Force Base fly through the fire area at an incredibly low altitude. I investigated quickly and called a very understanding colonel at their dispatch center in Barksdale, LA. He immediately posted a Notice to Airmen (NOTAM) so that no additional traffic, either civilian or military, would fly within 5 miles of the fire area or below 4000 feet.

TEMP	RH	WIND	FFMC	DC	ISI	BUI	KBDI	R
54	88	W 10	38.1	522.9	0.1	73.3	408.3	1.

Spruce	ROS=2	FL=1	Grass	ROS=1	FL=1

SEPTEMBER 15

Mike Paluda ordered a direct attack on the North Zone. Thirty-two miles of fire line were needed around the immediate fire edge. The objective was to hold the fire in place and prevent further spread. A large acreage of private property to the north and east was a significant concern. Lower temperatures, higher humidity, and a rainfall of 0.38 inches slowed the spread of the fire.

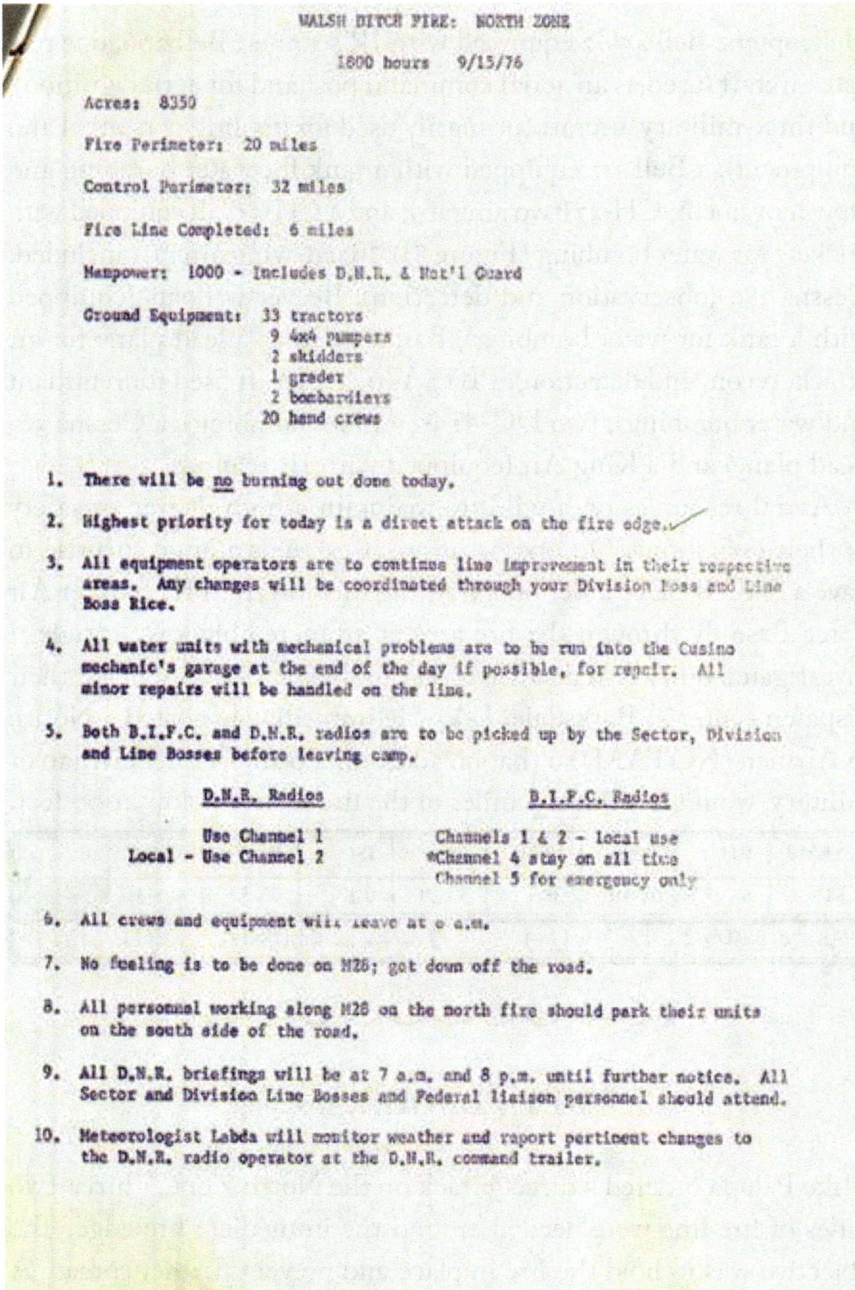

WALSH DITCH FIRE: NORTH ZONE

1800 hours 9/15/76

Acres: 8350

Fire Perimeter: 20 miles

Control Perimeter: 32 miles

Fire Line Completed: 6 miles

Manpower: 1000 - Includes D.N.R. & Nat'l Guard

Ground Equipment: 33 tractors
9 4x4 pumpers
2 skidders
1 grader
2 bombardiers
20 hand crews

1. There will be no burning out done today.

2. Highest priority for today is a direct attack on the fire edge.

3. All equipment operators are to continue line improvement in their respective areas. Any changes will be coordinated through your Division Boss and Line Boss Rice.

4. All water units with mechanical problems are to be run into the Casino mechanic's garage at the end of the day if possible, for repair. All minor repairs will be handled on the line.

5. Both B.I.F.C. and D.N.R. radios are to be picked up by the Sector, Division and Line Bosses before leaving camp.

D.N.R. Radios	B.I.F.C. Radios
Use Channel 1	Channels 1 & 2 - local use
Local - Use Channel 2	*Channel 4 stay on all time
	Channel 5 for emergency only

6. All crews and equipment will leave at 6 a.m.

7. No fueling is to be done on M28; get down off the road.

8. All personnel working along M28 on the north fire should park their units on the south side of the road.

9. All D.N.R. briefings will be at 7 a.m. and 8 p.m. until further notice. All Sector and Division Line Bosses and Federal liaison personnel should attend.

10. Meteorologist Labda will monitor weather and report pertinent changes to the D.N.R. radio operator at the D.N.R. command trailer.

Figure 83, A page from the daily fire plan for September 15, 1976

Resources on the fire line included 20 hand crews (400 firefighters on the fire line), 33 bulldozers (three with 36-inch wide tracks for higher flotation) and tractor-plows, two skidders with plow and water tanks, two bombardiers with water tanks, nine 4x4 pumper trucks and a grader (Figure 83).

TEMP	RH	WIND	FFMC	DC	ISI	BUI	KBDI	R
54	76	NE 5	55.9	526.8	0.5	74.1	411.7	L

Spruce	ROS=2	FL=1	Grass	ROS=1	FL=1

SEPTEMBER 17

A direct attack on the North Zone was mounted, with much of the fire line constructed by hand in the extremely difficult marsh and bog terrain.

Figure 84. Bureau of Land Management experimental fire truck, the Dragon Wagon

Several pieces of specialized equipment were brought in to aid in the suppression effort. The Bureau of Land Management (BLM) brought their large, experimental, eight-wheel-drive firetruck called the Dragon Wagon (Figure 83). It had torches mounted on both sides for igniting unburned fuel adjacent to the fire line, or it could spray fire retardant that was mixed onboard and applied directly as the vehicle moved along. Other items were a large muck-farm marsh plow, a helitack crew to cope with unique problems posed by this fire, and a giant sprinkler irrigation system, that used 200 sprinkler heads that could distribute a half-inch of water per hour each over a hundred-foot radius. The daily fire plan showed 1000 people on the North Fire, including MDNR and National Guard personnel. Ground equipment includes 30 tractors, nine 4x4 pumpers, three skidders, four bombardiers, and one grader.

TEMP	RH	WIND	FFMC	DC	ISI	BUI	KBDI	R
72	35	NE 3	87.5	537.7	3.8	80.5	414	M

Spruce	ROS=12	FL=12	Grass	ROS=9	FL=3

SEPTEMBER 18

The same combination of favorable burning conditions, warmer temperatures, low humidity, and moderate winds that previously caused the fire to become active and spread occurred again. The fire put intense pressure on the North Zone's east flank but was held within the lines thanks to aerial water drops and hard work by firefighters. The organization chart for the day is shown in Figure 85. A new fire, set by an arsonist, breaks out 15 miles to the southwest of the Seney fire and about four miles south of the Jack Pine Lodge in the Hiawatha National Forest. It "crowned out" quickly in the dry, mature jack pine. The fire was reported to our dispatch by aircraft on the Seney fire. I was able to contact two inbound loaded DC-4 air tankers and diverted them to this new fire. Their initial attack put the fire down

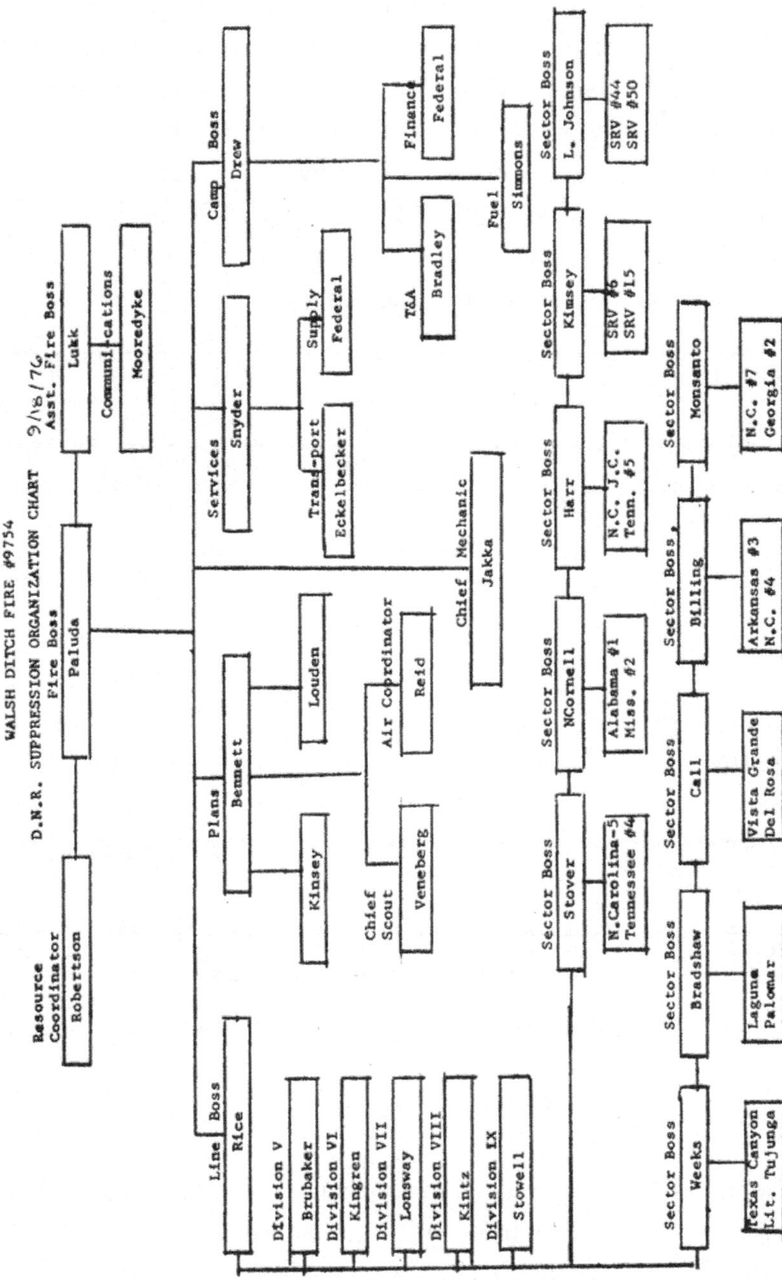

Figure 85. North Zone fire organization chart from daily fire plan, September 18, 1976

and allowed additional aerial and ground forces to help keep it relatively small when it was contained. Firefighters and equipment were diverted from the Seney Fire; with a tractor-plow from Manistique; at the height of the suppression activity, five D-4 air tankers, two large helicopters, and four 20-person hand crews contained this fire with a loss of 220 acres, saving the Jack Pine Lodge. A hunting camp was destroyed by the fire. Mop-up proved difficult; out-of-state Forest Service crews worked on it until early October.[90]

TEMP	RH	WIND	FFMC	DC	ISI	BUI	KBDI	R
72	57	W 4	87.5	543.4	4.1	83	416.3	H

Spruce	ROS=12	FL=12	Grass	ROS=9	FL=3

SEPTEMBER 20

Sporadic, light afternoon-long rains dampen the fire and hold it in place. The fire line was completed around the 32-mile North Zone perimeter.

TEMP	RH	WIND	FFMC	DC	ISI	BUI	KBDI	R
56	40	W 8	79	532	2	62.5	423.4	L

Spruce	ROS=4	FL=8	Grass	ROS=3	FL=2

SEPTEMBER 21

The Seney fire was again declared contained, mop up began. A loss of 8,500 acres in the North Zone (Figure 86) makes the total loss for the Seney Fire 72,500 acres (112 square miles), with 88 miles of control

90 Daily Press, Escanaba, MI, Threatening Fire Contained, September 20, 1976

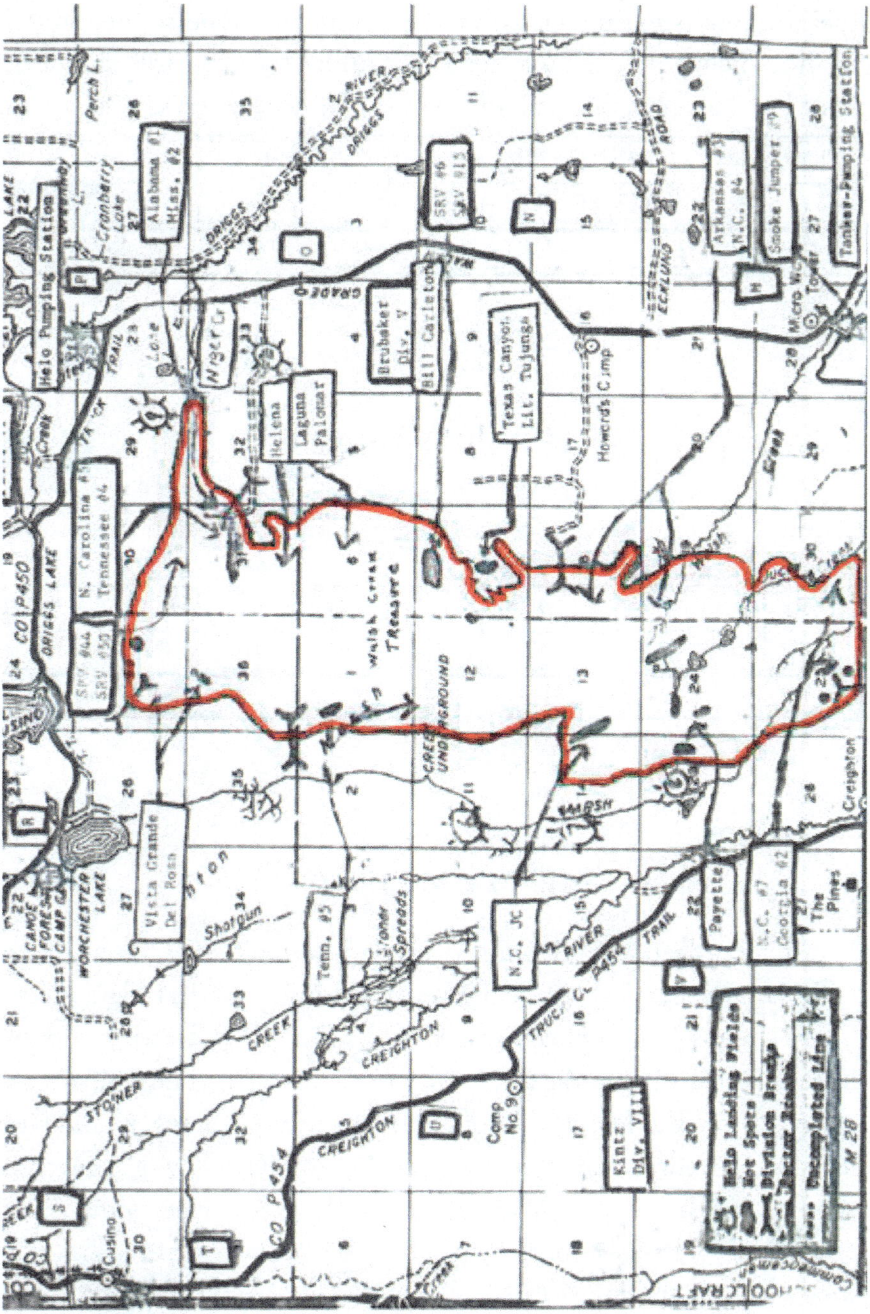

Figure 86. North Zone fire map from the daily plan, September 21, 1976

lines around its perimeter. Ground equipment and personnel on the fire line included 13 tractors, 14 4x4 pumper trucks, three skidders, three bombardiers, and 23 hand crews (460 firefighters).

TEMP	RH	WIND	FFMC	DC	ISI	BUI	KBDI	R
42	41	W 7	83.5	534.7	3	63.5	425.8	M

Spruce	ROS=7	FL=8	Grass	ROS=6	FL=3

SEPTEMBER 22

Full-scale mop-up began in the North Zone. Light rain with some snow was reported overnight.

TEMP	RH	WIND	FFMC	DC	ISI	BUI	KBDI	R
46	27	SW 12	87.9	537.8	8.3	65.2	431.6	H

Spruce	ROS=29	FL=12	Grass	ROS=27	FL=6

SEPTEMBER 23

A suspicion that several isolated fires in Schoolcraft County were deliberately set prompted cash rewards for information on arsonists. The Michigan State Police Fire Marshals Division offered $1,000 for information leading to the arrest and conviction of anyone deliberately setting a fire. The Jack Pine Lodge fire burned 220 acres of land and destroyed a hunting camp, and three other small fires were set within five miles of that area. A Michigan State Police sergeant pointed out that "there is plenty of motive for setting a fire because there's money to be made in putting out fires."[91]

91 Daily Press, Escanaba, MI, *"Arson Suspected,"* September 23, 1976

SEPTEMBER 24–30

Significant demobilization occurred; the fire transitioned to a Type II overhead team. Elmer Hurd was named the Federal Fire Boss. He and his Type II Bureau of Land Management team were from Alaska and had experience with fires in similar soils and vegetation.

Special equipment on standby included a 10-person helitack crew with a helicopter, sprinkler irrigation systems, and a DC-4 air tanker.

Approximately 300 firefighters remained on the fire and continued to mop-up. Thirty MDNR personnel were operating fifteen pieces of water pumping equipment. Forty-five support personnel, including the National Guard, provided food and transportation for the firefighters.

The North Zone was still hot in the center; the South Zone was extremely hot in some spots. Fire, burning in the organic ground fuels threatened the line in scattered locations.

A special team of investigators from the Michigan State Police Fire Marshals Division in Lansing and Marquette discovered six fires in the Seney vicinity. The fires were separate from the main fire and were believed to be set by arsonists. Investigators believed that some individuals were trying to maintain the need for equipment and workers, lining their pockets at the expense of the taxpayers through the destruction of the forests.

OCTOBER–DECEMBER

On October 10, the temperature dropped below freezing, and the first frost of the fall covered the fire area. On October 12, there was a report of 338 people still assigned to the fire, including 225 Federal firefighters, 85 state and private workers, and 28 National Guard personnel from Sault Ste. Marie.[92]

92 The Evening News, Sault Ste. Marie, MI, October 12, 1976

The last time the fire was flown was on November 3; no signs of smoke could be seen. At that time, the fire was covered with eight inches of snow. The Federal fire boss, Frank Sanders, classified the fire as abandoned. He had only two choices, that or controlled. The winter snowpack was the final and best mop-up for any remaining

Source: Walsh Ditch North Zone-Daily Fire Plan, September 6, 1976

Figure 87. Firefighters came from twenty nine states to fighte the Seney fire.

hotspots. The drought's effect carried through the winter and gradually abated by the following spring.[93]

FINAL ENTRY

Despite all the sophisticated equipment, any fire is, in the end, put out by people with tools, from shovels and pump cans to Dragon Wagons and helicopters. Twenty-two agencies provided Federal and state firefighters that came from 29 states representing all corners of the country, from Alaska and California to Georgia and Maine, to join the fight (Figure 87).

Many MDNR forest firefighters and specialists from the Forest Fire Division were pressed into service from all around the State

---

93 The Mining Journal, Marquette, MI, December 13, 1976, p. A-2

and performed admirably on the Seney fire, but being professionals in their field, this came as no surprise. The unknown factor in this critical situation was the many Department of Natural Resources staff that became involved that had never logged a single hour on a forest fire. They were Fisheries and Wildlife Biologists, Park Rangers, Foresters, Surveyors, Waterways, Geologists, and more than a few men and women who were normally anchored to a desk. Mechanics from around the State, normally surrounded by the conveniences of a repair shop, packed their toolboxes and performed heroics to keep the firefighting equipment in good working condition. The record of the MDNR employees that worked on the fire is only a partial list[94]; it is found in the Appendix.

Figure 88. Keith Owen, MDNR firefighter

It was exhausting work (Figure 88), but in retrospect, the Great Seney fire was a tribute to the men and women that pulled together, giving up their much more familiar positions to work on a common cause. The fire was contained through the combined efforts of the more than one thousand State and Federal firefighters, cooperators, and support staff that responded to the challenges of suppressing the largest fire in Michigan since 1908. Their remarkable effort stands out as an example of what dedicated professionals can do together as a team.

94 Department of Natural Resources, *Partial List of State Employees*, District 4 records, Newberry, MI October 11, 1976

CHAPTER SIX

HOTSHOTS' STORIES

A fire crew can have many adventures when they go on an assignment far from home. The Little Tujunga Hotshots from the Angeles National Forest published a newsletter in 2010 with recollections of their time in Michigan in the fall of 1976.[95] They were one of six Federal hotshot crews dispatched from California to assist in suppressing the Seney Fire. Following are their stories.

INFLIGHT EMERGENCY, 1976

It was raining in southern California, and the fire danger was down to zero. Walt Sniegowski was in the Little Tujunga Hotshot office when the telephone rang. The Forest dispatcher, Bob Underwood, asked to speak with him, and he immediately got this awful feeling in his gut. Could it be an early end to the fire season? After all, it was already September, and rumors of budget shortages were floating around.

Surprisingly, Bob asked if the crew would like to take a fire assignment, but it came with "conditions." Bob Underwood was a long-time fishing partner and a skilled practical joker. Both were fond of single malt whiskies and, on occasion, would enjoy one or two. He was sure that Bob might have already had several at lunch or breakfast, so his reply was full of expletives.

But he soon realized Bob was earnest. When asked about the "conditions," Bob stated the crew had to agree to stay for two weeks,

95 Sniegowski, Walt; Herrand, David; McGeragle, Mike; Vrandenberg, Mike; Satern, Monte; Wrench, Rod, Angeles National Forest, *Alumni Newsletter 40 Year Anniversary,* Edited and Arranged, Volume 1, No. 3, Summer 2010

leave all of their tools at home; and we were authorized administrative leave to go home and pack "essentials." The "essentials" were identified as raingear, rubber boots, a heavy jacket, long johns, and a warm sleeping bag. To go on a fire!

The fire was burning on the Seney National Wildlife Refuge in Michigan's Upper Peninsula, known by most as the UP. Named the Walsh Ditch Fire, it had been burning since July during a rare drought. Burning in what was usually a swamp, the fire was a rare extra-period incident (more than one shift). They needed help, so the US Forest Service, Region 5, would rotate 20-person crews for two-week shifts to supplement the State and Refuge crews.

A US Forest Service charter aircraft was ordered out of Boise. It would pick up Mark Linane and the Los Padres Shots at Goleta, CA; Larry Caplinger and Texas Canyon Shots and our crew at Burbank, CA; then Ron Regan and the Del Rosa Shots at Ontario, CA. Two Cleveland National Forest crews made other travel arrangements. To say the crews were excited would be an understatement. Making such a long trip is unusual, and we were looking forward to some fun, sharing stories and fellowship on board.

This was back when the Forest Service was beginning to get into "Total Mobility," recalled David Herrand. Going out of Region was not so unusual, but going as far as the Midwest was. So, we were excited to get this assignment. Plus, 1976 had been a relatively slow fire season for us, and our frustration deepened with significant rain in September. This fire assignment was sure to beat the project work we were doing up in the boneyard while wearing rain gear!

I remember the aircraft being a Convair 580 (Figure 89), a very good airplane with two big turboprop engines and a pressurized cabin capable of carrying some 45 or so passengers and their belongings. It was an ideal aircraft for carrying two 20-person hand crews and their gear.

We boarded at Hollywood Burbank with the Texas Canyon Hotshots. Our flight attendant wore a nametag identifying her as Bambi.

Figure 89. Convair 580

Bambi was an attractive, dishwater blond with big hair and single as far as we knew. I think all the California Shots and Jumpers knew of Bambi. Since most of us were single, it was inevitable that the guys would maneuver into talking with her. But we each learned that trying to speak with her beyond two or three lines was frustrating. It was not that Bambi was unfriendly. She was just confusing, and it was not clear to us if it was our fault or hers. After all, she was our flight attendant, and we looked up to her or thought we should.

We loaded our gear in the late afternoon and boarded the flight. The contract to fly us did not include feeding us anything more than one can of chilled soda pop per passenger, so Bambi didn't have a lot to do as far as food service. But we never left home without C-rations, so that is what we took for an in-flight meal. The first leg would take us to Denver, where we would stop for fuel, some impromptu stretching exercise, and a midnight candy bar.

David Herrand recalled never being able to sleep on an airplane. "I grabbed my usual window seat when we took off and resumed our flight east. It wasn't long before we started passing through a band of severe weather over the Great Plains. In the illumination provided by flashes of lightning, I watched as ice built up on the starboard wing. Repeatedly it would build up to a point, then suddenly, most of it would flake off. I was trying to figure out if it was flaking off from its weight or if the pilots were activating a de-icing boot when a massive flash of lightning bathed the wing in a brilliant, white light."

"I didn't overthink this and simply waited for the next flash. Then I saw the wing in a strangely dim light. The pilots had turned on an exterior light that shined out across the wing. There didn't appear to be much ice, but it looked like the propeller on that starboard engine had stopped spinning. I couldn't believe what I was seeing, rubbed my eyes, and looked again. Was I seeing things? Maybe I was dreaming. Then a flash of lightning confirmed it. The pilots must have shut that engine down! "Why?" I wondered. "Well, they will probably start it back up right away," I thought. Then the exterior light went off, and with each subsequent flash of lightning, I saw only a stationary prop."

I looked around the cabin at 39 other sleeping firefighters and one sleeping flight attendant in her unique seat way back in the tail. We all had our boots off, and most had removed their Nomex shirts and seat belts. I thought about walking back and waking Bambi. "Bad idea," I thought. Instead, I woke the firefighter next to me and told him that we were flying on one engine. Irritated to have been wakened, he opened his right eye, looked out the window, and mumbled, "Oh, they do that to save gas," and immediately went back to sleep.

I'm embarrassed to admit I pondered this for a few minutes. Searching the slide show in my mind, I couldn't explain why a two-engine airplane with over 40 souls on board was flying on one engine for this considerable length of time. I was an Army paratrooper for crying out loud! I had spent many dozens of hours cramped in four-engine turboprops or four-engine jets boring holes in night skies. Those Air Force boys never shut down any engines to save any stinking gas! The worst part was I didn't have a parachute on tonight!

Walt Sniegowski added to these troubling thoughts. Well, it was fun until about midway thru the flight! We were seated at the front of the plane, and most of the guys were catching a few Z's. One of our crew called over to me and said smoke was coming out of the right engine, and the propeller was not turning. Note that this is a two-engine airplane. Also waking up at this time was our flight attendant, Bambi. She proceeded to run, screaming and pounding on the cockpit door. We got a bit concerned.

David said it was past time for action. The crew foreman, Mike McGeragle, was seated directly in front of me. I woke him and told him what was happening and suggested we wake the crew and get them booted up. Shaking off his sleepiness, he immediately grasped the situation and agreed. We woke those within arm's reach and ordered, "Boots, Nomex, and seat belts, pass the word." One by one, the firefighters came to life. Grumbling sleepily, they nonetheless passed the word, turned on their reading lights, and slowly began to comply. Soon the Texas Canyon Shots were doing likewise, and Bambi was waking up.

Then the pilot came on the cabin speakers and asked Bambi to come up to the flight deck. She obediently strode up the aisle and disappeared behind the door to the nose of the airplane. Within moments she reappeared, looking shaken. Throwing her arms up, she shouted, "We're all going to die! Fasten your seat belts and put your heads between your knees!" With that, she ran pell-mell down the full length of the aisle, threw herself into her seat, strapped on her seat belt, grabbed a pillow into her lap, and buried her head into it!

Picture, if you can, 40 young men, now wide-eyed, each looking over their shoulder at Bambi with their jaws hanging open. Suddenly 80 boots were laced at light speed, Nomex shirts were donned in a fury, and 40 seat belts snapped tight almost in unison.

Walt remembered that a short while later, the pilot came into the cabin area and briefed the crews. He stated the engine had possibly been struck by lightning and was turned off to prevent a fire from occurring in it. He further informed us that the aircraft was entirely capable of flying on only one engine if necessary; that in about 15 minutes, we would be landing in Sioux City, Iowa. A replacement aircraft would meet us there, and we would continue on our way. He returned to the cockpit and closed the door.

David recalled that, in stunning contrast to Bambi's above-described "announcement," the pilot calmly announced over the cabin speakers that we had been hit by lightning and the alternator on number two had burned out. He said that the procedure was to shut the engine down and land at the closest suitable airport. He advised us we would proceed to and land, at Sioux Falls, Iowa, in another few minutes. While we were trying to figure out why we were over Iowa, indeed, if Iowa even had a Sioux Falls, he came back on and corrected himself- Sioux Falls, South Dakota!

Walt remembered that it took longer than 15 minutes; it was closer to one hour. We constantly shifted our attention from the good engine to our watches and back again during this time. The only sounds heard emanating from the crew members were an assortment of prayers and the rattling of beads. There were no atheists on board. David recalled that they flew on long enough to enter the first light of dawn. Descending, they eventually caught sight of the airport and began to fly a left-turning pattern around the area at about 3000' above ground. Our imaginations ran wild. Were we burning off fuel? Were we giving them time to foam the runway? Through all of this, I would occasionally look back at Bambi, only to see that her head was firmly buried in her pillow. "At least it was light now, and the severe weather had dissipated or moved on," I thought.

Walt added that the flaps were lowered on approach, and we landed. "I can recall that when we finally taxied up and read the airport name on the main terminal building, we were sure. It said Sioux Falls, South Dakota! Reflecting on my geography knowledge, I did not believe that Sioux City and Sioux Falls were that close together. I also asked myself why the main airport building was called "terminal," but I decided to let it go."

Finally, the pilot announced we were going to land. We descended slowly through some more orbits, then turned on final and made the smoothest landing I had ever experienced in a fixed-wing airplane! As anticlimactic as this was, Bambi then rose from her seat, combed her hair with her fingers, smoothed her dress, opened the door, and stood by as the mobile stairway was rolled into position! We got off the airplane and were standing around wondering if we should start unloading our gear when the pilot and co-pilot emerged from the aircraft and came down the stairway. The senior pilot was a mature and very distinguished-looking man. I will never forget how calm and collected he looked in his impeccable black coat and tie uniform. I will probably never know if those pilots simply messed with poor Bambi's head that night.

Because we had no landing permit for Sioux Falls, we were told the crews could not use the terminal's facilities. We considered passing the hat to purchase one, but instead, all crew members, with Buck knives at the ready, convinced the rent-a-cop/airport security guard that it was in his best interest to help us. Shortly we, or perhaps more so, our cash, was royally welcomed in the terminal, where we ate and drank all that was available. Several hours later, singing our crew song, we boarded the replacement aircraft and continued to the assignment.

A FIRE LIKE NO OTHER, WALSH DITCH-1976

The crews landed in Marquette in the Upper Peninsula (UP) of Michigan during the peak of the fall foliage season. The entire countryside was resplendent in shades of reds, oranges, and yellows. The fire was being run by a team from the Michigan Department of Natural Resources (MDNR). Although Michigan has a history of large fires, this generation of MDNR folks had never been tasked with a blaze of this size and duration. There was no pre-planned fire organization. Most of the crews on scene were state crews from the northeast and south on 30-day contracts. At the same time, an overhead team from the Boise Interagency Fire Center (BIFC) arrived to oversee management of the fire.

David said this thing was essentially a mega grass fire. Despite infrequent thundershowers and wet fronts, the UP was experiencing a severe drought. The fire was burning in an area that was usually open marsh and timbered swamps. The fire would rip through the dead grass that had grown on these dry marshes and wetlands; I mean, at times, it RIPPED!

Roaring in the open, the first tree stand it would hit was replacement conifers that stood on the ground about one foot higher than the grassland. It would crown big time through the conifers until it got into the hardwoods that stood on the ground another foot taller, called "ridges" by the locals. The fire would crown a bit into the hardwoods but quickly drop and become, first a running, then a smoldering ground fire. On these two-foot-high "ridges" in the shade of the hardwoods, we attempted to stop the thing.

Adapting to the local ways took a little time and patience. First, they only worked one shift a day. They don't fight fire at night because it's too dark, and they didn't have any "lanterns," headlamps to the rest of us. The daily briefings were experiences to remember. We met in a facility that we immediately nicknamed the "revival tent." Each morning the individual crew leaders and overhead shared their previous day's accomplishments, usually to a boisterous round of applause. It was not unlike "show and tell" back in grade school.

We would get a daily pep talk, safety briefing, and fire behavior prediction. We learned new fire behavior terminology: scorch height in the hardwoods, peat moisture content, and new measures of spread rates, furlongs per fortnight, I believe. Keep in mind that it was raining off and on most of the time, and very soon, snow flurries would begin-but the peat continued to burn.

David and Mike McGeragle both remembered what the locals called the northern peat. The called it loons—t, a spongy mixture of decaying vegetation and a hair-like root tangle, knitted together with a weird white membrane. One day we were mopping up with assistance from a couple of MDNR, one-person pumpers. These "engines" were large, military surplus pressure washers mounted on 1-ton trucks. These rigs carried maybe 200 gallons of water and 50 feet of a high-pressure hose with a nozzle that resembled a Thompson Machine Gun. The MDNR guys would cut the loons—t up with the thin, high-pressure straight stream; it was a hydraulic mop up on steroids.

The previous day we had constructed a direct handline through the hardwoods. Initially, our "California style" method involved a four-step process. First, we made a 10-foot-wide cut, which was the easy part due to the trees' wide spacing. The cut was followed with 3-foot wide, 6- to 10-inch-deep grubbing and scraping down to the spongy loons—t that was shot through with masses of tree roots. This was by far the heaviest scraping any of us had ever encountered. The fun part was then chopping and digging a trench one shovel width wide down to the water table. What we ended up with looked like an 18 to 24" deep meandering irrigation ditch. The fourth step was, of course, mop up.

We only did this labor-intensive method for one or two days because it didn't even begin to work. When we came back out the following morning, our "mopped-up" fire had crept over our ditch like it wasn't even there! So, what we accomplished was nothing more than a stupid little canal wandering aimlessly through the burn! The MDNR guys got a big laugh out of us California boys! I remember Mike and I asked one of them, "What is this stuff, and what does it take to put it out?" He reached down and ripped a handful of it up. Then said something like, "You boys think you can put loons—t out? Watch this." He rolled it into a ball in his hands and squeezed it, just like a sponge, until a drop or three of water dripped out. Then he fluffed it out and began blowing on it. No lie, with only a few puffs, it started smoking and sparking. A few more puffs and it erupted in open flame! We realized there were long fingers smoldering deep within the loons—t. It was impossible to find the edge of the fire. We adapted by minimizing line construction and instead mop lining. If you kept at it day after day, you would eventually get it. Of course, this was some two decades before handheld heat sensors. Because of the constant rain and snow predictions, it was necessary to construct what would eventually be known as "Visqueen City" (Figure 90). Huge rolls of black plastic were brought in for protective shelters.

Each crew was allotted time to erect its own camp. Armed with hand tools and chainsaws, we constructed the Michigan equivalent of the "swamp" from the television show M*A*S*H. Complete with directional arrows to tropical paradises (home), our camp contained fire rings, clothes, boot-drying racks, and benches around a community fire. The roof was sloped to allow runoff, something other crews failed to do. Proper drainage was installed, as well as adequate ventilation for the warming fires.

Using a couple of paper sleeping bags, several blankets, and a trash bag full of leaves as a pillow, we made quite the cozy lodging. The MDNR constructed a central dining area, but it had no seats. It didn't matter. It rained so often they wouldn't have gotten much use anyway. We ate, standing at chest-high plywood tables where you ate as fast as possible, shielding your paper plate from the constant rain. The UP was experiencing an extremely high jobless rate. An effort

Figure 90. "Visquean City", black plastic crew tents

was made to help the economy by hiring lots of local help. Some of them robbed the place blind, especially food and tools. If it wasn't nailed down, it was soon gone. Our crews were required to leave a man in camp each day to watch over the hotshots' property.

"The Walsh Ditch Fire, to me, was just a fun, fun fire," Mike remembered. We tried every type of construction method there was with the black plastic sheeting. Over time we figured out the only thing that worked was the tepee! It would rain every night; it rained every day! But, man, that fire would not go out! I remember them telling us when we first got there, "You boys with your fancy pants, shirts, and fire shelters, you ain't gonna put this fire out." Wait a minute! We're Hotshots of California; we put fire out! Two weeks later, we were reduced to whining, "We can't put this fire out!"

The Walsh Ditch Fire was a real eye-opener, and as for a unique adventure, it nearly pegged the needle. But the suppression work was incredibly labor-intensive and, frankly, by the second week, boring. As the days progressed, fatigue accumulated in the guys. You would think a fire on flat terrain would be easy. Our guys were in excellent shape and soldiered on like always, but they were fried in two weeks. The quality and the limited variety of the food didn't help. The cubic calories burned were more than the guys could replace. Fire camp was established in what was usually a swamp, so we named it "The Swamp"! The place looked like a ramshackle mining boomtown straight out of a bad Hollywood movie. We were directed to a far corner of The Swamp and told to make shelters for ourselves. The six R-5 crews dispersed into the woods, and each constructed a somewhat isolated little bedroom community. It was like a suburb to the main camp and became known as California or Little California, to no one's surprise. Camp construction and improvement never completely stopped. Local thieves prowled the camp, so we left a different person in our camp each day. Besides providing security, the in-camp person would improve the camp and restore the firewood supply. Our camp was a well-laid-out creation of relatively dry, snug shelters, multiple warming fire pits ringed by low benches, and drying racks after a few days. It became a real home away from home. It helped morale to come off shift each evening, wet and tired, to crackling warming fires. Had we stayed another two weeks, I think we would have even finished the log cabin we started.

David recalled a memorable event that occurred on one particularly heavy, rainy day. An alert was sent out over the crew radios for everyone to return to the fire camp about mid-morning. It was raining so hard that the overhead team feared the crew vehicles would get mired down in the swamp. Upon arrival in camp, I joined the crew for lunch while Mike McGeragle went to timekeeping to close out our workday. When asked by the MDNR timekeeper why we stopped

Figure 91. Little Tujunga Hot Shots-1976

the crew's time at noon, Mike pointed to the storm raging outside the tent rather than the usual six pm. Her remark, "Not your fault it rained"! They paid us for the full day. Mike added: I remember we came in early because it was raining cats and dogs, and the creeks were rising. I stopped our time at noon. The timekeeper asked me, "Well, what have you been normally working"? I said, "Oh, 12 to 15 hours". "Then put that down. It's not your fault it rained," she replied!

Mike said: On one particularly cold day, the Service Chief came out on the fire line to see if our lunches (locally made) were adequate.

I made a joking remark that hot soup and coffee might be an excellent addition. Starting the next day, and for as long as we remained there, thermoses of hot soup and coffee showed up as regularly as clockwork. Any suggestions for needed equipment or supplies at the daily shift briefings quickly became a reality. Mark Linane asked for and got fire line explosives. He was soon blasting fire lines through the swamp and "holes" where helicopters with Simms buckets dipped out water to make their drops. Others asked for and received sky cranes, dragon wagons, Bombardiers, muck plows, etc. The cost never seemed to be a concern.

A classic for me, said Mike, was the MDNR guy, probably the Line Boss or something. He was out on the line with us and called in a fixed-wing drop. Radioing the aircraft, he said, I'm waving my orange hard hat" I mean, the whole country is orange with fall leaves! I laughed so hard I thought I was going to die! But they had reason to laugh at us too. The first day we, all six Region 5 crews, I think, were sent out to cut a "shaded" fuel break, through the hardwoods, along a backcountry road. We cut 60 or 70 feet wide for well over a mile and did a really good job. Late that day, one of the good old boys showed up and said, "You dumb idiots from California. The only trees you left are the paper birch, which is the hardwood tree that burns the best!"

A lot of those DNR guys had never seen a fixed-wing retardant drop. I remember this one guy, probably the same guy that waved his orange hardhat, called in a drop. We were trying to figure out what he was dropping on. After the drop, one of us, I think it was me, asked him, "What in the world are you dropping on?" He said, "I just had to see me one of them retardant drops; it sure was the purtiest thing I have ever seen!" It got boring out there in the second week, not from lack of work; it was the monotony and futility of it. But being hotshots, we kept after it. Some of those out-of-state crews were horrible. I remember one crew came out there and slept. They didn't even break open their tool bundles.

At the end of two weeks, we were told they wanted two crews to stay and four would be released. Straws were drawn. I don't remember which two crews drew the short straws. The crews remaining would move into cabins, something like a Boy Scout or church camp. We helped get the cabins outfitted.

On the last day, Monte added, we set up some barrack space for the remaining crews. We hauled mattresses and beds from a truck into the rooms and assembled them. Then we were told that they were for the other crews and that we were to go back to our camp and sleep on the ground in our plastic tents again. Upon hearing that news, one of our guys fell to the ground and began to flop around in the dirt and babble about loving to sleep in the dirt. Our host was not impressed.

We ate one meal at this new camp Mike said. Different people prepared it, and it was way better than what we had been eating for two weeks. All during my hotshot career, I have always felt that our crews have received (and probably deserved) a bit more "attention" than others have. But the Michigan experience was like no other (if you could discount the weather). Our reputations certainly preceded us. The performance rating and letter from our Division Supervisor, Howard Weeks, and the entire BIFC Team were very special. We almost hated to leave, but that would be a small lie. I use the word "almost" because four of the six crews would be released when it was time to go. We would draw straws. The two "losing" crews would move from "Visquean City" to winter quarters in the local schoolhouse. They would remain for an additional week. As the bright lights of Manistique, Escanaba, and Munising faded from the cabin windows deep inside, I was honestly quite glad that I had drawn the "long" straw.

What did a Forest Service firefighter earn on a big fire in 1976? The Little Tujunga Hotshots (Figure 91) were assigned to the Walsh Ditch Fire for the full 14 days of one pay period, September 12 thru Saturday, September 25, 1976, with two days off during the pay period. One firefighter's time report for the period showed gross earnings for a GS-4 step 1, with regular pay, hazard pay, Sunday pay, and overtime of $1292.67. A GS-3-1 would have received $1163.40; a GS-5 step 1-$1421.94; a GS-6 step 1-$1564.13.

In 2022, the Infrastructure Investment and Jobs Act had a provision to increase Federal wildland firefighter compensation as part of a classification and compensation reform package. The law intended to increase firefighter pay by 50% or $20k for all primary and secondary firefighters regardless of GS level. For example, under the law, a GS-3 earning $13.78 in 2022 would make $20.67 an hour for base pay and a roughly $31.00 overtime rate.[96]

96 Wildfire Today, *Federal firefighter classification without compensation-wait, what?*, https://wildfiretoday.com/2022/06/16/federal-wildland-firefighter-classification-without-compensation-wait-what/, June 16, 2022

CHAPTER SEVEN

THE FINAL TALLY

CONTAINED, CONTROLLED, OUT

The fire was contained on September 21. When fire authorities describe a blaze as "contained," they do not imply that the fire is "controlled" or that it is extinguished. For a fire to be considered "out," it must first be classified as "contained" and then "controlled." Thus "controlled" is, one step closer to a fire being considered "out" or fully extinguished. When a fire is "contained," a control line has been completed around the fire, and any associated spot fires, that can reasonably be expected to stop the fire's spread. The fire can continue to burn actively within the bounds of the line, but its spread is less likely. "Contained" means that a barrier has been built around the fire's perimeter that is expected to stop the fire from spreading. "Controlled" implies that a control line has been constructed, and any hotspots or unburnt fuel near the line have been removed from the immediate vicinity. For a fire to be considered "out," no hotspots can be detected within the containment lines for at least 48 hours.

The highest Drought Code (576.1) and Keetch-Byram Drought Index (450) were reached on October 4. Cool and wet conditions set in, and the fire danger was low to moderate until winter arrived, but it was still possible the fire was burning in remote locations deep in the swamp.

The fire was abandoned when heavy winter snow finally blanketed the Upper Peninsula. The following spring, the heavy winter snowfalls thawed, finally extinguishing any remaining smoldering organic soils that burned as deep as eight feet beneath the surface.

The Great Seney Fire was only then declared out. There were worries that it could continue burning into 1977; peat fires in Michigan have burned for several years.

FINAL SIZE

The fire was estimated to be 72,500 acres, over 110 square miles, with 88 miles of perimeter. In 1977 and 1996, other sources reported the final fire size at 74,000 to 76,000 acres. In 1996 Fish and Wildlife Service officials stated that of the total acreage reported initially, only 35,623 acres were burned within the original perimeter.

INJURIES

Despite the potential dangers, there were no significant injuries. There were scrapes, bumps, bruises, and a few broken bones. The Marquette Office of Emergency Services coordinated the paramedics and five emergency medical technicians assigned to the fire. They reported over one thousand visits during their first 2 ½ weeks; most were for cold remedies. The most serious medical problem was not fire-related; a firefighter was brought to the hospital in Munising by helicopter with an appendicitis attack.

TIMBER DAMAGE

Written MDNR interoffice communication from Wesley A. Lehmann, Upper Peninsula Regional Forest Supervisor, to Glenn M. Schaap, Section Leader, Forest Cultivation & State Forest Operations, Lansing, dated April 27, 1977, indicated the timber loss and regeneration costs submitted for collection to the Federal government was $161,148.67. Recalculating this using 2021 stumpage rates gives a cost of $805,051. In 1976, Cleveland Cliffs, which owned much of the private land burned, claimed their losses were $400,000; accounting for inflation, this cost would be almost $2.1 million in 2022.

SUPPRESSION COSTS

The final cost estimates for fire suppression were over $8,000,000 with over $1,000,000 to the State, plus timber damage and regeneration costs. The total suppression cost in 2022 would be almost $40 million, adjusted for inflation. Of that amount, the cost to the MDNR, adjusted for inflation, would be nearly $5 million for suppression, plus approximately $800,000 in timber damage and forest regeneration costs on State land.

During the suppression of this fire, many other fires were occurring throughout the Upper Peninsula; some of them had the potential to become large, problem fires. An indirect effect of the Seney fire was the lack of skilled staff to suppress these other fires since most full-time MDNR firefighters were on the Seney fire. This may have caused increased costs of suppressing these other fires, an added cost to the State.

FLORA AND FAUNA

The Walsh Ditch Fire provided a rare opportunity to investigate the effect of a landscape-scale fire on the Refuge. To assess the initial impact of the fire on the land and its wildlife, the Seney Fire Evaluation Team was formed. At the direction of the Fish and Wildlife Service, a 3-year study was initiated in the spring of 1977 detailing the effects of the fire. Dr. Stanley H. Anderson of the Migratory Bird and Habitat Research Laboratory in Laurel, Maryland, was designated Project Leader and was assisted by several other biologists. The fire's ecological effects given here are taken from Dr. Anderson's report. The team was tasked with three main goals: 1) to evaluate how much the fire had impacted fish, wildlife, and their habitats, 2) to provide a baseline for future monitoring of changes in the environment, and 3) to compile a document publicizing the findings and provide recommendations for the future. The impact of the fire on the wildlife was studied using seven 8-hectare paired (burned-unburned) plots. The vegetation responses varied across the burned plots due to variability in fire severity. In many areas, the effects of the fire were hardly noticeable after the second growing season. Plots closer to the Walsh Ditch showed more altered conditions relative to the unburned plots. A final report detailing the effects of the fire on wildlife and wildlife habitat was released in 1982.[97]

The fire burned in irregular patterns with varying degrees of intensity. In some burned areas, dead trees or downed logs were the only vegetative remains, whereas others were difficult to recognize as burned areas after leaves covered the ground. The patchy nature of the fire caused the diversity of wildlife to increase on the Refuge. Where the fire was severe, structural changes occurred in the habitat creating areas for species such as sharp-tailed grouse, black-backed three-toed woodpeckers, and black bears. American woodcock and snipe, already common on the Refuge, became more abundant as the "edge" habitat increased. White-tailed deer showed little change because of the lack of winter habitat. Studies in the third year after the fire indicated that young amphibians and reptiles were more common on the burned plots than on the unburned study sites. Fish populations were similar in ponds and streams near burned and unburned regions. Significant differences were not found in the chemical composition of water and soils at burned or unburned sites. The overall impact of the fire was to create a more diverse habitat by setting back succession in some areas.[98]

A post-fire vegetative cover map (Figure 92) prepared by researchers in this study was compared to pre-fire vegetation to determine changes that occurred because of the fire. This data was created from an interpretation of Landsat satellite images in 1977. It shows the degree of burn on a relative scale (0 indicates a complete burn and 4 a light burn). Estimates made by the research team showed that of the total acreage

97 Anderson, Stanley H., Resource Publication 146 July 1982, *Effects of the 1976 Seney National Wildlife Refuge Wildfire on Wildlife and Wildlife Habitats*, Fish and Wildlife Service U.S. Department of the Interior

98 Anderson, Resource Publication 146

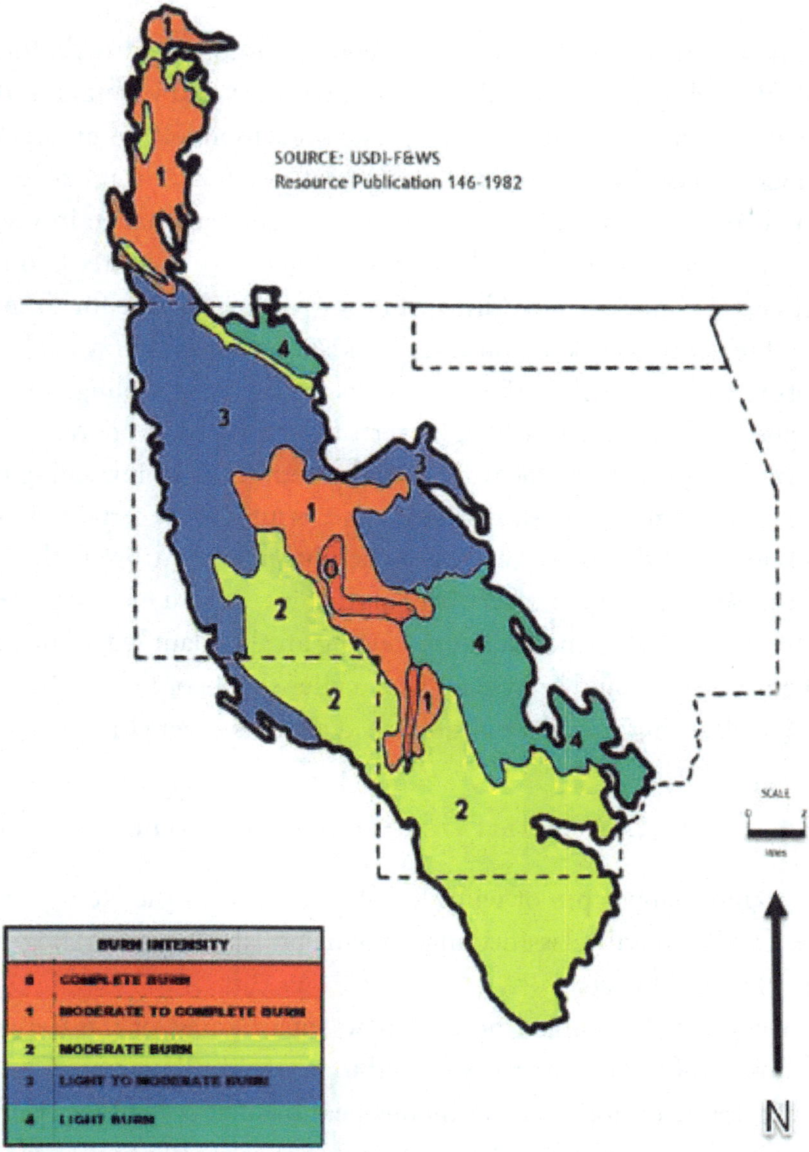

Figure 92. Burn severity map created from satellite photos in 1977

within the fire perimeter, the fire burned, to varying degrees, a total of 35,623 acres (37%).[99]

---

99 Anderson, Resource Publication 146

HOW DID FIRE AFFECT THE WILDLIFE?

Each animal and plant need specific conditions for survival. Animals are affected by other animals, diseases, weather, and vegetation. All these influences change constantly. The vegetation follows an orderly change (succession) where one group of plants makes it possible for a new group to thrive and dominate. This gradual change in vegetation, where grassland yields to shrubs, then shrubs to trees, forces out certain species of wildlife while providing habitat for others. Fire breaks this cycle by pushing back succession. After a wildfire, mature trees are killed, the site reverts back to tree saplings, brush sprouts, or grass, and brushland reverts to brush sprouts or grass. This change, in turn, affects the wildlife in the area. Animals inhabiting the area during earlier plant successional stages return, while certain species present before the fire leave because their habitat is destroyed. The fire on the Seney Refuge was very irregular. This created a mosaic pattern of burned and unburned areas, producing an abundant "edge" habitat where animals could browse on new tender shoots or feed on berries and seeds, using the adjacent unburned areas as cover (Figure 93).[100]

WILDLIFE HABITATS ON THE SENEY REFUGE

Although many types of wildlife habitats exist on the Refuge, the area can be broadly classified into two major habitat types: 1) bog-fen and 2) mixed forest.

The bog-fen occupies more than 20 square miles of the Refuge. It consists of treeless string bogs and topographically oriented strips of bog forest on top of a continuous peat blanket several feet thick, covering a layer of sand. The water table is relatively high, usually just below the ground's surface. This vast open land is interrupted by many knolls or islands of sand that project above the peat blanket and range in size from 1/10 of an acre to several acres. Red pine,

100 Anderson, Resource Publication 146

Figure 93. Images of the burned area in 1976, showing a mosaic pattern

white pine, or birch trees grow on these knolls. The open bog area consists of low bog ridges about 12 to 15 feet apart, alternating with wet sedgy hollows. Many of the ridges have small tamaracks growing on them. Projecting between the open bog areas are fingers of bog forest growing on the peat. These forests are composed of tamarack, black spruce, red maples, and an occasional birch. These landforms are more highly developed in the northern latitudes of Alaska, Canada, and northern Europe. The center of the Refuge is one of the southernmost extensions of this unique habitat.[101]

HABITAT CHANGES FOLLOWING THE FIRE

Vegetation structure was the major wildlife habitat component affected by fire. These structural changes were most pronounced in the bog-fen area. Here the fire burned deeply into some of the bog, creating large caverns. The treeless bog strips received deep patterns of burn. Trees died in the bog forest and on sand islands as the fire destroyed the shallow root systems. Some bog forests were virtually burned to the ground, and the trees remained standing or fell haphazardly in other areas. Grasses invaded most of the bum area by the third growing season following the fire.

The fire burned erratically in the mixed forest area, destroying most of the vegetation in some areas but only some in others as it burned around tree bases. Quaking aspen and jack pine seedlings, as well as blueberry and bracken fern, grew on the burned ground in the spring following the fire (1977). By the spring of 1978, two years after the fire, much of the ground had a rich cover of aspen seedlings and later in the summer large patches of blueberry.[102]

Biologists sampled the water in and near the burned area during May 1977 and 1978. Their results showed that no measurable mineral

101-102 Anderson, Resource Publication 146

characteristics threatened fish or wildlife. Even though much of the cover vegetation was destroyed, the very thick humus-like material throughout the Refuge held minerals rather than releasing them into the water.

WILDLIFE CHANGES FOLLOWING THE FIRE

Because the 1976 fire did not burn a continuous block, most wildlife found refuge in the many scattered areas not touched by the fire. Research on wildlife species in the three years following the fire indicated that the number of species increased. Generally, the burned areas attracted larger numbers of small mammals, bird species and more predatory animals.[103]

Fish population samples, including northern pike, yellow perch, and brown bullhead, were taken on lakes and streams immediately adjacent to or within the burned areas and in streams: several miles away. No significant differences were found in these populations in the two years following the fire. This was attributed to the ability of the peat to hold minerals and silt leaving the mineral content of the water relatively unchanged.[104]

Although cape may warblers, mourning warblers, hermit thrushes, swamp sparrows, and ruby-crowned kinglets were displaced because of the loss of dense thickets and forests for nesting, habitat for new species of birds was created. Brown creepers and eastern kingbirds increased in numbers. The black-backed three-toed and other woodpeckers and flycatchers became relatively common in the area. In the burned bog forest, the bird population was reduced to only a few species, including the yellowthroat, white-throated sparrow, lincoln's sparrow, song sparrow, and brown thrasher. The two species found in the burned but not the unburned areas were the brown thrasher and the song sparrow.

103-104 Anderson, Resource Publication 146

The fire destroyed the prime marsh waterfowl nesting habitat near some of the dikes, thus reducing some populations of nesting waterfowl in the years immediately following the fire. This was somewhat balanced by the newly created grassy edges of the burned bog, which provided new nesting habitat for ring-necked ducks, mallards, black ducks, and green and blue-winged teals. Immediately following the fire, migratory Canada geese were seen feeding on the burned fields. The greater sandhill crane, which typically nests in open areas with dense ground cover near shallow marshes, benefited from the fire. Even though some bog-fen nesting habitat was destroyed, the lush grasses that appeared near water impoundments in the year following the fire created additional areas for this species to nest. The fire opened up some habitat for sharp-tailed grouse, which were then seen occasionally in open areas between the mixed forest stands and the grasses growing on the burned bog-fen.[105]

The fire helped to maintain the woodcock habitat and increased the woodcock population. The common snipe population also appeared very high in the third spring following the fire. Snipe were commonly flushed along the boggy marsh edges throughout the Refuge.

Small mammal populations decreased. Overall, the density of mammals was very low in the spring of 1977. By October, one year after the fire, small mammals increased 17-fold on burned sites. This change had a significant impact on predators seeking food. Small mammals, such as shrews and voles, were quite common in the late spring of 1977. Alder thickets that grew in the third year following the fire were used extensively by beavers. The small aspen seedlings began to appear, providing food for beaver, deer, and small mammals.[106]

White-tailed deer, common animals in much of the Upper Peninsula of Michigan, are not found in large numbers on the Refuge primarily because of a lack of protective winter habitat; deep snows

---

105-106 Anderson, Resource Publication 146

force deer to congregate in areas with adequate food and cover. So, even though shoots and grassy vegetation increased due to the fire providing ideal summer habitat, the deer population did not seem to benefit because of a lack of suitable winter habitat. The dense crops of berries that followed the fire provided ideal feeding areas for black bears.

HABITAT CHANGES FROM THE FIRE

The overall effect of the fire was to create a greater diversity of wildlife habitat on the Refuge by setting back natural succession. This change included eliminating understory and the appearance of snags, downed logs, and edge. Vegetation structure was the major wildlife habitat component affected by fire and was altered most drastically in the peatland areas of the Refuge. On the treeless bog strips, burn scars will remain for years. The bog forests also showed considerable change since most of this cover type was burned to the ground. On many of these sites, aspen, jack pine, and blueberry became established in a relatively short period.[107]

Without additional fires, habitats and wildlife populations will approach pre-fire conditions. The humus that supports trees in the bog-fen took hundreds of years to accumulate. Mosses and lichens will help stabilize the exposed sand. Grasses will then become established; however, the many shallow-rooted trees will not reappear for years.

The 2008 Refuge wildland fire plan states that "In the future, fire should be used to create and maintain quality early successional habitat within both terrestrial and wetland ecosystems by reducing woody vegetation cover. Fire will promote breeding habitat comprised of grasses and sedges for a wide range of wetland obligate species (e.g., waterfowl, yellow rail, sedge wren (*Cistothorus palustris*), Le Conte's sparrow (*Ammodramus leconteii*). Within xeric terrestrial ecosystems, fire and other silvicultural tools can be used to maintain

107 Anderson, Resource Publication 146

habitat for other open land species of conservation concern (e.g., sharp-tailed grouse, upland sandpiper). In red and white pine ecosystems, fire use may promote savanna-like ecosystems, whereas in jack pine dominated sites, the use of fire should promote a patchy distribution of regenerating jack pine and open lands. Such conditions may favor species such as Kirtland's warbler (*Dendroica kirtlandii*) as well as a number of winter finches that are in global decline. Fire can also promote vigorous aspen regeneration where desired."[108]

POLITICS AND PUBLIC OPINION

Media sources in the region thought they saw a lack of good judgment by Federal officials. Following is an example of the sentiment at the time, as expressed in an editorial in The Mining Journal dated August 25, 1976, entitled "Investigation Required."[109]

> *Officials at the Seney National Wildlife Refuge made a big mistake when they did not allow Michigan Department of Natural Resources firefighters to extinguish a lightning-caused forest fire on the Refuge before it became a major conflagration and threatened State and private lands.*
>
> *The situation is a good example of what can happen when Federal policy is so rigid it precludes-or intimidates-reaction to exigencies by workers in the field.*
>
> *The Seney fire, which has already destroyed more than 18,000 acres of forest and is still raging out of control, was discovered as a small blaze over three weeks ago. It was not extinguished because it was in a "designated wilderness area" where mechanized equipment is forbidden and "a little fire is a good thing," according to Federal policy.*
>
> *We don't take issue with the policy as such but with the apparent*

---

108 *Wildland Fire Management Plan, Seney National Wildlife Refuge*, Department of Interior, Fish and Wildlife Service, 2008

109 The Mining Journal, Marquette, MI, *"Investigation Required,"* p. A-8, August 25, 1976

inflexibility of its application.

The Upper Peninsula is presently suffering one of the worst droughts in recent memory. Wells are running dry, streams have turned into trickles, and the entire forest is crying for a good drink. Spot fires have been cropping up like mushrooms after a spring rain.

Under those conditions, even the most unsophisticated observer could have predicted that a "little" fire would turn into a big one.

There should be an in-depth investigation of the circumstances surrounding the Seney fire to determine who allowed it to get out of control. If it was an error in judgment by a local official or due to the arrogance of some bureaucrat in Washington, he should be fired. If it was the inflexibility of policy governing wilderness areas, it must be changed, and the person responsible for it relegated to counting the bodies of dead animals and charred trees scarring the Upper Peninsula landscape.

Another report in the Mining Journal on August 25, 1976, continued similarly. It said, "Michigan Department of Natural Resources officials were furious with a Federal policy that allowed the Refuge to burn for days without attempts to douse it and that they were planning to bill the United States government an estimated $100,000 in State damages, equipment, and firefighting labor. 'Whether we'd go as far as a court fight would depend on their response to our request,' Wayne Tody, DNR Deputy Director, said this morning." He added that "no serious discussion of cost and damage recovery would take place with Department of Interior officials until after the blaze was brought under control."[110]

On August 31, U.S. Representative Philip Ruppe, R-Houghton, sent a telegram to Assistant Secretary of the Interior, Nathaniel Reed, inviting him to accompany Ruppe on a tour of the fire area asking him to order more equipment and personnel to the Seney blaze to assist in battling the fire. Ruppe also said he would seek Congressional hearings on the prescribed burning policy (Figure 94). Ruppe said,

110 The Mining Journal, Marquette, MI, *"DNR to bill Uncle Sam,"* August 25, 1976 "

"It has been suggested that prescribed burning is...a... tool of wilderness management, but that is absolute folly when, as in this case, it threatens other significant forests, wildlife or recreation resources."[111]

By September 4, Congressman Ruppe said that the House Merchant Marine and Fisheries Subcommittee on Fish and Wildlife had "..agreed to his request for a congressional hearing on the managing of the fire at the Seney Wildlife Refuge in the Upper Peninsula." Ruppe said the purpose of the hearing was twofold. "The U.S. Fish and Wildlife Service policy of controlled burning in wilderness areas is in need of a thorough review. My reading of the regulations gives the agency much greater latitude in fighting the Seney fire than has been exercised. Either poor judgment was involved in response to the fire, or the agency is not interpreting its guidelines for fighting fires the way I believe Congress intended." Ruppe also said, "There is the question of reimbursing the State for the cost it has incurred fighting the fire." Lynn Greenwalt, Director of the Fish and Wildlife Service, a few days earlier, while on an inspection tour of the fire, had stated his belief that the Interior Department should pick up the tab for the State's expenses.[112]

Also reported on September 4, Jack Hemphill, Regional Director of the U.S. Fish and Wildlife Service, said the fire was not initially ignored as several State officials have claimed. He said no order was issued to let the fire burn. Hemphill said that Refuge personnel attempted to extinguish the fire within two days after its location was pinpointed. He added that the Federal code of regulations forbids the use of the "let burn" policy during dry weather, such as Michigan was experiencing.[113]

On September 16, Jack Frye told a congressional hearing that Federal policy kept him from sending mechanized equipment into the Refuge's

---

111 The Mining Journal, Marquette, MI, *"Seney fire near end"*, August 31, 1976

112 The Mining Journal, Marquette, MI, *"Congressional hearing,"* September 4, 1976

113 Escanaba Daily Press, *"Fire hearing slated,"* September 4, 1976

Figure 94. There was considerable political fallout from the Seney fire

"dedicated wilderness area.." Frye said, "I felt I could not fight the fire with mechanized equipment because the area was under a wilderness designation," and that he felt constrained until he conferred with his Regional Office and was given the go-ahead. This contradicted previous statements from Federal officials about wilderness policy. Nathaniel Reed, assistant secretary of the interior, and Lynn Greenwalt, director of the Fish and Wildlife Service, told the committee that there is no "let burn" policy for wilderness areas. Howard Tanner, Director of the Michigan Department of Natural Resources, said the DNR's policy was to extinguish the fire immediately regardless of wilderness policy. Greenwalt stated that the Department of Interior would pay all costs associated with fighting the fire. One official said this was the first time the Federal government said it would pick up the State's costs. Greenwalt added that any reimbursement to private concerns must be settled through court action. This remark came after testimony from Cleveland Cliffs Iron Company said that their company had lost $400,000 worth of timber burned north of M-28.[114]

Interior Secretary Thomas Kleppe told 70 community leaders at a Marquette meeting on September 23 that the total cost estimate of fighting the fire was $4.1 million, of which $1 million was spent by the State. "We're going to pay it…the expenses the State has incurred in fighting this fire." At a meeting later that day, he said there was no "let burn" policy concerning the Seney fire, as Refuge Manager Jack Frye had claimed. "There was not a conscious policy of saying let it burn...If there is any blame to be assessed, we ought to probably be blamed for miscalculation, misunderstanding, or a lack of total knowledge for the kind of fire it was."[115]

A few days later, in a press conference in Dearborn, MI, Kleppe said his department would review its handling of the fire, including the controversial "let burn" rule that State officials blamed for letting the blaze get out of control. The Refuge Manager admitted to letting the blaze burn for nearly two weeks after it was started by lightning on July 31. " We didn't just let it burn," Kleppe said. But he admitted there was no major effort to control the fire for several weeks; "It shouldn't have gotten to the level it did." Kleppe said that no disciplinary action was planned, and he wasn't aware of any bad judgment in handling the fire, but an internal review of the matter was intended to answer those questions.[116]

An editorial in the Marquette Mining Journal on November 12 tied the November 9 death of Smokey Bear, the living symbol of forest fire prevention for 25 years at the National Zoo in Washington, D.C., to a broken heart when he was forced to sit by and watch the destruction of 106 square miles in the Seney Wildlife Refuge because of the Federal "let burn" policy. "A hearing into the 'let burn' policy

---

114 The Mining Journal, Marquette, MI, *"Let burn hassle rages on,"* September 16, 1976

115 The Mining Journal, Marquette, MI, *"Sec Kleppe confirms feds will pay for Seney fire"*, p. 3-A, September 23, 1976

116 he Mining Journal, Marquette, MI, *"Seney fire handling to be reviewed; Kleppe will study 'let-burn' policy"*, September 23, 1976

governing wilderness areas has been held. It was scheduled prior to the November 2 election to allow for maximum political exposure. All that apparently came out of the hearing, however, was a bunch of mumbo-jumbo, doubletalk, and buck-passing. Smokey is dead, but life goes on for the bureaucrats."[117]

The Federal government reimbursed the State of Michigan by early the following June. It was reported on June 7, 1977 that the State Department of Natural Resources told Gov. William Milliken that $919,814 was paid to the State's general fund by the Department of Interior. The governor had asked the Federal government for the money the previous September. Representative Ruppe's efforts helped secure the reimbursement.[118]

Reports the following summer showed the political pressure from Representative Ruppe had not abated. It was reported that Ruppe would conduct a hearing to evaluate the interior department's fire-fighting policy for the Seney Wildlife Refuge on June 28, 1977. He had demanded the completion of a fire plan for the Refuge by March; that deadline came and went when he said, "The interior department has had more than substantial time to complete a thorough, effective policy. I am interested in seeing what significant changes have been made in last year's fire plan." It was also reported that a letter from Ruppe to Fish and Wildlife Secretary Robert Herbst urged expediency in developing the Seney plan. Ruppe added, "we hope to see this plan confront special circumstances such as last year's drought conditions which transformed Seney into a forest of fuel. The hearing will investigate and evaluate the procedure for reimbursement of both private landowners affected by the fire and the State, as well as the fire plan itself."[119]

Compounding the confusion over fire policy, the Federal

---

117 The Mining Journal, Marquette, MI, *"A broken heart,"* November 12, 1976

118 Escanaba Daily Press, *"Feds Pay State Nearly $1 Million for Seney Fire"*, June 7, 1977

119 Escanaba Daily Press, *"Seney hearing set,"* June 24, 1977

government, also in 1976, shifted its fiscal year from July to October, leaving a vacuum during which it was unclear whether the Refuge even had money to pay for suppression. It was unsure where to turn for help until a phone call to the national office suggested BIFC, which in the absence of fire activity elsewhere, had funneled the national fire cache into Seney. Without lines of authority and clear rules of engagement, everything got exaggerated: the power of the burns, their damages, the chaotic response, and the political blustering. A review board later exonerated Refuge Manager Frye as having acted prudently "under the circumstances."[120]

FIRE MANAGEMENT ON NATIONAL WILDLIFE REFUGES

> Stephen Pyne wrote that: *the environmental movement had helped consolidate the Refuges into a system and had boosted the agency's responsibility for treaties governing migratory birds. It began as the Bureau of Sport Fisheries and Wildlife (known to its sister agencies in Interior as "Fin and Fur") and as overseer of the 1969 Endangered Species Act, which then upgraded in 1973 into the most potent piece of environmental legislation from the era. That same year the Convention on the International Traffic in Endangered Species, which the agency would also oversee, opened for signature; in 1974, it was renamed the Fish and Wildlife Service.*
>
> *But it was for fire a time of breakdown and even shame. The agency lacked for fire management the institutional contexts it had acquired for endangered species and wildlife Refuges. What it did right it could not promulgate throughout the system. What went wrong affected everyone. It should have been their time to shine. If the instinct of the BLM were to stop fires, that of the F&WS was to start them. The Refuges had by far the greatest experience in prescribed burning of any Federal agency. Refuges have utilized prescribed burning with the application of scientifically derived prescribed fire plans for habitat management*

---

120 Pyne, Stephen J., *Between Two Fires*, A Fire History of Contemporary America, The University of Arizona Press, Tucson, The Arizona Board of Regents, 2015

and wildlife protection for many years.

The F&WS had officially utilized controlled fire since the early 1930s. Refuge workers, if not their managers, tended to come from the adjacent populace; they understood local culture and, where burning persisted, local practices. But if it had the hands-on, one-foot-in-the-black knowledge that most of the national agencies lacked, it lacked the institutional knowledge. It had no organic act that defined the place of fire. It had no national office to supervise fire operations —to assist technology transfer, to disperse the latest science, to inform practitioners about policy reforms, or to adapt the National Fire Danger Rating System to their usage. It lacked the ability to train to national standards or to represent the agency in the National Wildfire Coordinating Group.

Prior to 1978, the Refuge staff did not have the necessary fire suppression training and experience to make critical operational decisions beyond the initial attack of wildfires. Since they did not have a formal fire control organization, fire suppression responsibility was contracted to other Federal agencies or the State of Michigan.

During and after the fire, accusations swirled like embers in a stiff wind. The Refuge staff had not been tested by a fire of any kind in five years, it did not have experience in prescribed burning, and it lacked fire-danger rating capabilities. The Refuges outdated fire plan did not include the mapped wilderness, and it had no national guidance. It simply could not cope, certainly not with two problem burns. The Refuge was uncertain what it could do with a wilderness.[121]

FIRE REVIEW

The Fish and Wildlife Service convened an official review board, as requested by the Department of Interior, Assistant Secretary Jack Horton. During the period September 28-October 1, 1976 the Walsh Ditch Fire Board of Review met at the Ramada Inn, Manistique, Michigan. The Board of Review reviewed extensive information compiled by the

121 Pyne, Stephen J., *Between Two Fires,*

firefighting personnel assigned to the line. The Board of Review was comprised of high-ranking staff from the Fish and Wildlife Service, National Park Service, Michigan Department of Natural Resources, Bureau of Land Management and U.S. Forest Service; it was chaired by Jim Pulliam of the Fish and Wildlife Service. The board gathered information from agency staff and personnel that managed the fire; fifteen personnel were interviewed.

The board spent one day on the fire, flying it by helicopter first and then receiving a briefing from Fire Boss Elmer Hurd and the overhead team. The remainder of the week was spent interviewing key personnel and preparing their official findings. After reviewing and examining the information they gathered, they believed a comprehensive review had been accomplished.

During the interviews, it became apparent to the board that two critical decisions were made that significantly affected the course of the fire. These two decisions were (1) not to enter the wilderness initially and (2) the withdrawal order of firefighting resources due to uncertain funding capabilities.

The board noted that an additional key factor was the extreme drought experienced in the Upper Peninsula which was not readily apparent to the Refuge staff. The drought, coupled with temperature and humidity abnormalities, resulted in dangerous fuel conditions on the Refuge. This was compounded by the staff's lack of fire experience, the last significant fire occurring in 1970. Had they been more experienced, they would undoubtedly been more conscious of the building fire hazard potential. The board found, in part, that:

> *"As a result of this review, it is clearly apparent to the Review Board that the F&WS is now and must continue to be dependent on outside assistance when major fire control problems occur. The Service should strengthen initial attack capabilities on Refuges, however, and it is imperative that local and regional protection agreements be developed with cooperating fire protection agencies."*

The board also felt that the Boise Interagency Fire Center's availability to offer support was a crucial factor in suppressing the fire and that the recent expansion of BIFC's role in the east was fortunate. The board was impressed by the professionalism of the support groups and fire teams. They were also greatly impressed by the highly professional and cooperative assistance the State agencies and local entities provided. "Despite major complexities and the potential for great misunderstandings, State personnel worked effectively and harmoniously with Federal personnel."[122]

122 Pulliam, Jim, Chairman, *Report of Walsh Ditch Fire Board of Review,* October 1, 1976

CHAPTER EIGHT

PROGRESS

RADIO COMMUNICATIONS

The "new" technology of the 1940s may sound improbable to us today. When I started with the Forest Fire Division in 1972, I heard the story of a firefighter who had used one of the first radio-equipped fire vehicles in the old Michigan Department of Conversation. On an early spring fire in the late 1940s, a seasoned fire officer from Baraga was dispatched to a smoke on the Peshekee Grade. He had been blessed to operate the district's first pumper truck with a two-way radio. By this time, Conservation Officers had radios in personal cars used on patrol. His dislike and distrust of this new technology did not work in his favor. He refused to use it. The pilot in the spotter aircraft was trying to talk to him and direct him to the fire, and it took him many hours longer than was necessary to find the fire on his own and then put it out. Following an after-action debriefing and more training, he eventually came to rely on it as a necessary tool. The use of radio communications during a wildfire is a given today.

The early MDNR radio network was low-band; four frequencies around 44.64 MHz were used statewide for administrative and tactical purposes. For practical purposes, only two frequencies were utilized in the Upper Peninsula. The Seney NWR also used the same low-band frequencies. Other Federal agencies, especially the Forest Service, used high-band radios, so when they and State agencies worked together, it was always difficult to communicate by radio.

At the 1976 Seney fire, this problem was resolved relatively well by using high-band radios from the BIFC cache, operating on six channels around 168.050 MHz; some were assigned to key MDNR personnel and equipment. A separate frequency used only by the local fire departments and subsidized by the MDNR was 154.295 MHz. It was initially supposed to be a meet-and-greet frequency, not used for tactical purposes, although it was often pressed into that role.

A new low point for radio communications in the MDNR occurred on May 6, 1986, "Black Tuesday," as it came to be remembered by many central UP firefighters. It was a series of fires in the Upper Peninsula, probably more than 100, most in Marquette County. Early in the day, we knew there would be problems; it was very windy and dry—only 0.06" of rain had fallen, making it the driest May on record. There were predictions of a low-level touch-down of the jet stream; this was confirmed later in the day with winds over 60 miles per hour at times. Most of the fires were caused by downed power lines, with a few by arsonists. Probably the most problematic was the Sporley Lake fire burning in jack pine near K.I. Sawyer Air Force Base. The 4,200 acre, mostly crown fire, caused the evacuation of the 8,000 base residents and the commander's extreme concern about the safety of the nuclear weapons storage area. Extreme measures taken by firefighters, local fire departments, equipment operators, and military personnel successfully kept it from the base. Local fire departments handled the majority of the fires. Twenty-four of the larger problem fires were handled by the MDNR with assistance from cooperators across the Upper Peninsula. Of these, four fires totaled almost 8,000 acres lost. Several homes, camps, vehicles, and personal property were lost, but it could have been much worse. Perhaps the weakest link in the system was radio communications. With 60-70 or more radios in use on four large fires within range of one another, operating on basically one frequency, it was amazing the efforts were not completely tied in knots. There was confusion and delays, but the system worked because everyone was patient and restricted their traffic to essentials. We were fortunate no one got hurt that day.

Few remain today in the firefighting ranks of the MDNR that can remember those days and the vast improvement that has taken place in the State communications system. A bit of history about the evolution of that system might be valuable.

The development of the present 800 MHz system started in 1984 when a multi-agency State committee evaluated its 1940's era two-way radio system. The committee recommended building a new system that would be large and flexible enough to support all State and local public safety agencies. In 1992, after several years of system design planning, a proposal was sent to potential vendors. In June 1994, the Michigan Legislature approved the funding for the new system. Motorola was awarded the contract to build the Michigan Public Safety Communications System (MPSCS). The network was completed, in several phases, starting in the lower peninsula in 1995 and finally in the Upper Peninsula in 2002. It was first implemented to serve the Michigan State Police, then adding the Departments of Natural Resources and Transportation that have now become reliant on the system. The MPSCS is state of the art; it handles the second-largest trunked communication system globally, including administering a statewide 800/700 MHz digital trunked radio communication network. As of March 2020, the system spans 59,415 square miles, has 279 radio towers, and includes more than 108,000 radios.[123] Although they usually use high-band radios, Federal agencies also have 800 MHz radios and can cooperate with the State system as necessary.

LFO VERSUS ICS

In 1976 the MDNR Forest Fire Division still used the Large Fire Organization (LFO). The basic LFO organizational structure consisted of a Fire Boss supervising a Line Boss and a Plans Chief. The Fire Boss was the modern-day Incident Commander (IC), while the Line Boss and Plans Chief were the modern-day Operations Section Chief and Planning Section Chief, respectively. While several areas of LFO proved inadequate to complex incident management demands, other

123 *MPSCS-Michigan's Public Safety Communications System*, https://www.michigan.gov/mpscs/inside accessed June 10, 2021

components worked well. The Plans function of LFO had several similarities with the modern-day Incident Command System Planning Section and the old LFO Plans. Under LFO Plans, the Intelligence Officer and the Maps – Records Officer performed much of the modern-day Situation Unit functions. Similarly, the Status – Check-in Officer performed modern-day Resources Unit functions. While the Maps – Records Officer position has gone away and the function has been absorbed into the Situation Unit and Documentation Unit, the Status – Check-in Officer lives on as the Status Check-in Recorder in the Resources Unit.

As an incident management system, LFO could expand to incorporate multiple agencies, but its downfall was that it lacked a strong central coordinating mechanism.

The Incident Command System (ICS) is a systematic tool used to command, control, and coordinate any emergency response. ICS allows agencies to work together using common terminology and operating procedures for controlling personnel, facilities, equipment, and communications at a single incident scene.

While the intent in 1974 was for ICS to be used to manage all wildland fire field activities for the fire service, the system's design intent almost immediately evolved into an all-risk, all-hazard system. Early in the development process, participants recognized that the system they were designing needed to be simple enough for the "common man," work within each department's day-to-day response structure, and be effective for a wide range of incidents, simple to complex. Having different systems for different types of incidents did not make sense. A member of the development Task Force noted that they "had to build a system that worked for a dumpster fire, a high-rise fire, a flood, or a major hazmat incident."

The Incident Command System (ICS) is flexible, scaling up or down as complexity and needs change. Type 5 is the least complex, while Type 1 is the most complex. Area command is established when an incident is so large that it must be divided and managed as two or more separate incidents; or when multiple, separate incidents with Incident Management Teams (IMT) must be managed. The role of area command is to provide oversight direction to multiple incidents rather than giving direct action on any one incident as a Type I or Type II IMT would. Area Command manages the efforts of various Incident Commanders to ensure that the overall objectives are being met, to set priorities among incidents, and to allocate scarce resources between incidents. The LFO organization used the GHQ, General Headquarters, the early equivalent of Area Command, on the Seney Fire.

Early in 1976, the Department of Interior created six Type 1 overhead teams for national interagency use. One of these teams, from BIFC, with MDNR assistance, finally suppressed the fire utilizing many costly techniques. Until then, almost no one had heard of, much less used Unified or Joint command on an incident. Of importance is the fact that it was a cooperative effort. The Federal and State fire bosses (Incident Commander's) decided to share command and merge the line and planning functions jointly. Again, the concept was not familiar to anyone. By the end of the fire, State and Federal employees were serving in nearly all positions interchangeably. The credit for plowing fresh ground needs to go to these early pioneers, and of course, unified command is now utilized routinely all over the nation.

NATIONWIDE COOPERATION

Over the last twenty-five years, the levels of cooperation between Federal and State agencies have reached new levels. The Eastern Area Coordination Center (EACC)[124] now serves Federal and State wildland fire agencies within the twenty-state Eastern Area (E.A.). EACC provides logistical and resource support, predictive services, and intelligence for anticipated and ongoing wildland fire activity. EACC facilitates the movement of resources (people, aircraft, ground

124 *Eastern Area Coordination Center*, https://gacc.nifc.gov/eacc/admin/about_us/about_us.htm, accessed June 11, 2021

equipment) among the EACC member agencies and their units. EACC monitors wildfire potential, weather, and wildland fire use within the E.A. EACC also responds to requests for support to other geographic areas from the National Interagency Coordination Center (NICC) at Boise, ID. The United States and Alaska are divided into eleven Geographic Areas for incident management and mobilization of resources (people, aircraft, ground equipment). An interagency Geographic Area Coordinating Group (GACG)[125], made up of Fire Directors from each Federal and State land management agency within the area, is established within each area. Working collaboratively, the GACG's mission is to provide leadership and support for wildland fire emergencies and other emergency incidents (i.e., earthquakes, floods, hurricanes, tornadoes, etc.) as necessary. The authority for establishing the GACG is through departmental policy and interagency agreements. Additional agreements are established with cooperators and other organizations to facilitate efficient fire management activities within and adjacent to the area. Such cost-effective resource-sharing among public agencies is a critical component of the GACG mission and is expected by the public, Congress, and States.

EVOLUTION OF THE F&WS

Currently, the Seney NWR is restoring portions of the landscape to pre-European settlement conditions based on published patterns and processes. Prescribed fires are conducted yearly; however, it is unknown whether or not these "controlled fires" reflect natural wildfire patterns or if specific habitats are being targeted more frequently than one would expect from a random occurrence.[126]

After decades of persistence, prescribed fires have become integral to the Refuge's land management practices. Still, in 1996, the Refuge

---

125 *Geographic Area Coordination Centers*, https://gacc.nifc.gov/admin/disclaimer/disclaimer.htm, accessed June 11, 2021

126 Marsh, Dawn S., *The Evolution of Land and Fire Management at Seney National Wildlife Refuge: From Game to Ecosystem Management*, December 2013

underwent budget cuts, and the prescribed fire program's budget was reduced from $125,000 to $20,000, and the number of prescribed fires conducted declined significantly. Although the prescribed fire program had been reduced, the Refuge focused on restoring regional-level biodiversity levels by restoring and managing forest and wetland composition and structure. In 2003, the primary fire management focus was to rebuild the prescribed fire program after it had been absent for nearly five years. The following year, the Seney NWR conducted 14 prescribed fires. The largest prescribed fire since 1947 was conducted in 2009 and burned approximately 4,500 acres.

In 1978 the F&WS Fire Management Program was formally instituted when the position of Fire Management Coordinator was filled, making the F&WS Fire Management Program the youngest of the five Federal land management agencies. In 1979, F&WS joined the other Federal agencies in headquartering its fire program at the National Interagency Fire Center (NIFC) in Boise, Idaho.[127]

In 1988, more than 1.5 million acres of Refuge lands were burned by wildfire in Alaska, although this event was overshadowed by the highly publicized fires within the greater Yellowstone area. The F&WS's current fire management and budget reflect increased safety and workload needs, with about 500 full-time fire employees and more than 2,000 F&WS employees qualified ("red-carded") to support wildland fire incidents. By 1983, the national F&WS fire management budget was $4.5 million, with 64 full-time fire employees. But due to the 1988 fire season, new support by Congress for fire management resulted in an F&WS fire budget increase to $16.4 million, with 214 full-time fire staff. In addition, the Secretaries of Agriculture and the Interior evaluated existing fire management policies and recommended actions to address problems experienced during the 1988 season. The 1994 South Canyon Fire, 1998 Florida Fire Complex, 1999 Northern Nevada Fire Complex, and 2000 Western Fire Complexes culminated

127 *U.S. Fish & Wildlife Service, Fire Management,* http://www.fws.gov/fire/who_we_are/history.shtml, last updated May 28, 2014

in reviewing the Federal fire policy and creating the National Fire Plan.

30TH ANNIVERSARY

Bob Webber, fire boss on the Seney fire, sent an email[128] to the Seney NWR manager on the 30th anniversary (Figure 95) of the Walsh Ditch Fire; there was a get together of firefighters at the Refuge as a remembrance and review of what happened. His thoughts provide an excellent review of many of the significant changes that occurred as a result of the fire:

> *The Department of Interior created six class one overhead teams in 1976 for National Interagency use. Our team was one of the six. No one had ever heard of Unified or Joint Command. Therefore, when Mike (Paluda) and I decided that he and I would jointly share command and merge the line and planning functions, the concept was unfamiliar to most, even our teams. By the end of the tour, Federal and State Employees served in most functions. The credit for plowing fresh ground needs to go to these early pioneers; unified command is practiced routinely all over the nation now. I feel that there were many firsts in addition to Unified Command:*
>
> *1. Federal acceptance of State training and qualifications.*
> *2. Cross-nation use of personnel and equipment.*
> *3. Massive support efforts from almost every state.*
> *4. The recognition that every major fire belongs to all, regardless of where it occurs.*
> *5. The recognition that every major fire is of national significance.*
> *6. The recognition that all wildland and structural departments are one.*
> *7. A forever tie between east and west, state, military, private, and Federal forces.*
> *8. The establishment of the Fish and Wildlife Service as a fire agency.*
> *9. Standardization of equipment, training, emergency management organization, policies, etc.*
> *10. Representative Ruppe's committee in Washington called John Russell*

128 *Webber, Bob, email to Tracy Casselman,* Seney NWR Manager, September 2006

and me on the carpet for actions taken. Fortunately, it turned out to be primarily election-year politics.

11. Many good fire and management people were generated on this fire. Their experience was passed down to the next generation.

12. The fire will always be credited for its role in interagency cooperation.

13. This fire firmly highlighted the need for an organization like the National Wildfire Coordination Group and its working teams.

It was my pleasure to have served on several of those groups over the last twenty years of my career. I can't think of another fire ever that changed more attitudes, policies, and interagency procedures.

Figure 95. 30th Seney Fire Reunion

CHAPTER NINE

THE FUTURE

WEATHER AND CLIMATE

According to recent studies, rising temperatures and drier weather will increase across the continent and wildfires could double in the Great Lakes region.[129] Scientists predict that, with warmer temperatures and rising atmospheric CO_2, forest productivity, evapotranspiration, and lightning activity will also increase. These changes could affect all forest types and increase the risk of more severe summer wildfires throughout the region. Forests in the Great Lakes region are projected to experience increased insect outbreaks and drought, leading to increased tree mortality and fuel accumulation. Increased precipitation in the fall and spring may contribute to vegetation growth that would dry out and die during hotter, long summer days without rain, providing the setting for more severe wildfires. Drier conditions would bring an increase in the number and intensity of wildfires. Combined with an increased human presence in fire-prone areas, a changing climate will make wildfire prevention and control a continuing and increasing challenge.[130]

Wildfires in the U.S. have caused more and more damage over the years. More than 10.3 million acres were burned due to roughly 58,950 wildfires in 2020 alone. Five of the six largest wildfires in California history burned in August and September 2020, prompting a state of emergency declaration. The largest, the August Complex Fire,

129 Brock, Hannah, Great Lakes Echo, *Great Lakes wildfires could double,* https://greatlakesecho.org/2021/10/19/great-lakes-wildfires-could-double/, October 19, 2021

130 Neumann, Brad, C*limate Change and Wildfire in the Great Lakes Region*, Michigan State University, Extension Bulletin E-3277, August 2015

originated as 38 separate fires started by lightning strikes on August 16–17, 2020, lasting until November, burning 1.03 million acres. With a continued increase in temperature, wildfires are expected to remain a significant risk for decades to come. The U.S. National Climate Assessment predicted that the number of wildfires "is projected to increase further as the climate warms." The scientists predict that the total area burned will increase between 25% and 53% by 2100.[131]

In this century, rising temperatures in southern and western parts of the United States will cause climate migration to the north, which is already larger by some measures than political or economic migration. Because of the increase in the number of people, this migration will increase risk in the Great Lakes region.[132]

MANAGEMENT CHANGES

Fire managers face many challenges. Fires that start in extreme conditions must be professionally managed by well-trained, qualified firefighters using the latest equipment and technology. The increased use of prescribed fire could mitigate larger fires through fuel reduction, but some will escape and become large problematic fires.

To meet these challenges, wildfire agencies must move from short-term fixes to transformational change by enhancing firefighting capacity, including training, renewing critical assets, maintaining or replacing essential infrastructure, and developing innovative solutions to overcome human resource challenges. Complex incidents, other than wildfire will also increase in frequency and intensity. Increased cooperation at all levels will become necessary to bear a significant change in present climate trends. Of course, cooperation between partners will become even more critical.

---

131 Amadeo, Kimberly, *How Wildfires Impact the Economy*, https://www.thebalance.com/wildfires-economic-impact-4160764, accessed June 11, 2021

132 Yoder, Kate, Grist, *Fleeing global warming? 'Climate havens' aren't ready for you yet*, https://grist.org/migration/fleeing-global-warming-climate-havens-arent-ready-for-you-yet/, December 7, 2021, accessed April 16, 2022

The transformational change needed within wildfire management organizations to confront a rapidly changing world will be challenging. As wildfire events become more complex and more decision support tools become available, enhanced training and education will be required to strengthen wildfire management capability. Local, State, and Federal agencies must understand their authorized roles and responsibilities during complex, multi-agency wildfire emergencies. This component of wildfire preparedness will require agencies to foster and support a partnership culture before rather than during a disaster.

Increased preparedness capacity will be essential for the future by enhancing firefighting capacity, including training; renewing critical assets (aircraft, equipment, and highly qualified wildfire management personnel); maintaining or replacing critical infrastructure (communication and weather station networks, initial attack and air tanker bases); and developing innovative solutions to overcome human resource challenges (staff recruitment, retention, and skills upgrading). This includes understanding the authorized roles and responsibilities of local, State, and Federal agencies during complex, multi-agency wildfire emergencies. This component of wildfire preparedness will require agencies to foster and support a partnership culture before rather than during a disaster.

LINE BUILDING EQUIPMENT

Michigan doesn't have million-acre fires as it did in the 1800s because of mechanization and a road system to respond quickly with off-road firefighting equipment operated by skilled firefighters. In 1928 the Michigan Conservation Commission authorized the purchase of 12 trucks for fire control work, chiefly for hauling firefighters, supplies, and tools. At about this time, light trailers were provided for transporting supplies and equipment, and all forest officers' cars were fitted with trailer hitches. Trucks and trailers became standard equipment. The use of four-wheel drive trucks increased because of their better traction and ability to be used off-road.

State firefighting strategy in Michigan soon centered on the use of plows. Early efforts with horse-drawn plows demonstrated their usefulness for fire line construction. They were replaced quickly with similarly designed plows attached directly to a tractor with a hydraulic lift, which were first developed in the 1940s. By the early 1950s, the Michigan Forest Fire Experiment Station developed the first hydraulically operated fire line plow. It has been continually improved incrementally over the past decades to its present configuration. The latest iterations include environmental cabs on the tractor (Figure 97). The most recent has an updated plow system, including a rear-facing camera that shows the operator the plow and the fire line they are creating (Figures 96-98). Perhaps some future generation of

Figure 96. John Deere 450 tractor-plow plowing line

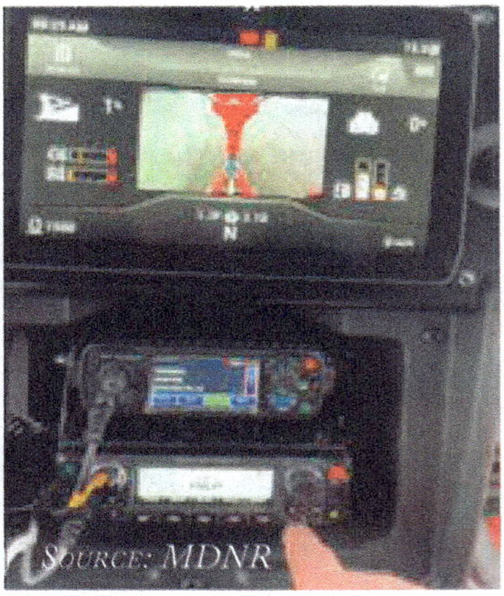

Figure 98. Cat D-4 tractor-plow, rear camera and radio

Figure 97. Caterpillar D-4 LGP tractor-plow with environmental cab

tractor-plows in Michigan will incorporate artificial intelligence and robotic controls and work autonomously.

The development of robotic equipment in the country now ranges from contraptions the size of a toy wagon to two-ton beasts that resemble military tanks and can blast out 2,500 gallons of water per minute. At present most are being developed for structural fire usage. Some move on rubber tires, some on steel tracks, and some fly. All are robots. At a time when more than 3,000 Americans die in fires each year—including an average of 80 firefighters—these high-tech devices can enter burning buildings too hot for human survival. They can penetrate smoke too toxic for human lungs. They are often faster, stronger, and more agile than the firefighters they work with. Most of the machines currently in use are remote-controlled, but researchers are developing "intelligent" firefighting robots that can make decisions autonomously. Machines will not wholly replace humans on the fire line. They are tools, and one valuable thing they can do is reduce risks to human firefighters. With more than 350,000 American homes burning annually, and climate change contributing to 10,000 active daily wildfires worldwide, robots can offer some respite to firefighters entering burning buildings or fighting a dangerous wildfire in rugged terrain. The main challenge these machines face will be an institutional reluctance to invest in devices tailored to meet these varying niche needs.[133] Researchers are working to change this. Hopefully, future development extends into the creation of autonomous line-building equipment.

AMPHIBIOUS VEHICLES

Future amphibious technologies, whether they are remotely operated vehicles, autonomous agents, assistive devices, or novel control interfaces, offer many promising capabilities for deployment in real-world environments. Many disaster events, including wildfires, present the

133 Little, Jane Braxton, *Firefighting Robots Go Autonomous*, Scientific American, October 29, 2021

need for rescue operations in aquatic environments. Beyond the need for ground and aerial robots to be simply resistant to weather or adverse conditions, amphibious robots will require significant engineering to enable aquatic operation.

Comparing the suppression strategy and tactics used on the Pine Creek North Wildfire that started on May 21, 2012, and the 1976 Seney fire shows that significant improvements have been made by the F&WS and the MDNR in preparation for wetlands fires.

While drought conditions were not a factor in the Pine Creek North Wildfire, as they were in the 1976 Seney fire, the weather conditions were quite conducive to burning in late May, 2012. In 1976 there were no trained firefighters on staff or amphibious equipment available to dispatch. By 2012 a key member of the Refuge staff was a Fire Management Officer (FMO) who had spent many hours preparing for just such an emergency. The Refuge had established fire management plans that assisted in managing an incident such as this. With plans in hand, the FMO and other trained Refuge firefighters, were able to effectively call for cooperators and resources and begin to suppress the fire. No one had to try to walk into the fire like the Refuge Manager attempted to do in 1976. They were equipped with excellent amphibious equipment and trained, qualified staff.

On May 23, the fire made a run that threw burning embers a distance of quarter mile across Pine Creek Road and Pine Creek igniting a spot fire. This occurred before the crews had time to complete a burn out operation along Pine Creek Road. The spot fire burned 400 acres before it was finally contained.[134]

Seven amphibious tracked vehicles, mostly Marshmasters (Figure 99), two CL-215 air tankers and a large helicopter were used to help slow and eventually put out this fire. An effective technique used when dealing with this portion of the fire was to use compressed lines where the amphibious vehicles drove over marsh plants and pressed

134 *Pine Creek North Wildfire-Seney National Wildlife Refuge*, https: //www.fws.gov/fieldnotes/regmap.cfm?arskey=32575

them into the water to help slow the fire's spread. These compressed lines were then burned out to create a barrier.

Unified command of all agencies was utilized with an overhead team consisting of U.S. Fish and Wildlife Service, Forest Service, and Park Service personnel and the MDNR's fire management team. About one and a half weeks after the wildfire began it was listed as contained. In the end, roughly 3,420 acres burned. It might be said

Figure 99. Marshmaster at MDNR Seney Field Office

that preparation for suppression of this wildfire started in 1976 and improved through the years with improved fire planning, acquiring and training fire staff and firefighters, equipment acquisition, and improved cooperation with other agencies. It is reasonable to assume that continued improvements will be made in the future.

SPOTTING FIRES FASTER

Well over one hundred years ago, with no towers or telephones available, it is easy to see why the early fire wardens were able to accomplish so little actual forest fire suppression. Locating fires by patrol was like looking for a needle in a haystack and it was only by accident that they were discovered while small enough to handle. The first lookout towers were trees or high points used by the patrolmen to obtain a wider view of the country. A tall tree near Raco, in the eastern Upper Peninsula, was long used for this purpose. The use of such vantage points led to the erection of small windmill-type towers with open platforms at high points along patrol routes. The result was the early substitution of lookout towers for patrol. The first lookout tower in Michigan was a low wooden affair built in the vicinity of Lewiston about 1912. The first steel tower was erected by the State on the Houghton Lake State Forest in 1913. In 1915, the second steel tower was erected on the Hanson Game Refuge near Grayling. The following year three towers were ordered for the Upper Peninsula.

Figure 100. Michigan Department of Conservation fire tower, 1940's

About this time towers with enclosed cabins began to be used in place of the old open platform type and stairways began to replace ladders. By 1928 the tower system was considered practically complete and most of the primary towers were of the enclosed cabin type (Figure 100). In 1949, the State had 140 primary towers[135]. Most towers were erected

---

135 Mitchell, J.A., and Robson, D., *Forest Fires and Forest Fire Control in Michigan*, Michigan Department of Conservation, In Cooperation with United States Department of Agriculture Forest Service, Lake States Forest Experiment Station, with the University of Minnesota, University Farm, St. Paul, MN, 1950, p. 36

on high points to increase their range of vision. The average effective range of such towers was about 10 miles or less in some directions depending on the topography. When visibility was poor, secondary towers were staffed in emergencies, aerial patrol was provided by State or hired aircraft.

Fire towers are no longer used in Michigan for wildfire detection. For the last fifty years or so, fire detection has been accomplished through the coordinated, preplanned, use of State and Federal aircraft and reports from the public and other agencies. Typically, the fires found by State and Federal aircraft are those in the most remote areas, and are the most problematic. On average, nine out of ten fires in Michigan are caused by people and most often they are relatively accessible. Lightning fires that occur in drought or extreme fire conditions have the potential to become large and difficult to suppress. These types of fires are most often found by aerial detection.

Soon, we may be able to leverage satellite and emerging technologies to rapidly detect new, often remote, wildland fires. The Federal government has several new, innovative ways to detect and notify emergency personnel when wildfires start. The National Oceanic and Atmospheric Administration (NOAA) uses satellite-based early wildfire detection to provide initial location information to first responders, allowing them to attack fires more quickly. NOAA is partnering with other Federal departments and State agencies to expand the use of this technology and its associated notification system. The Department of Homeland Security (DHS) is also focusing its research agenda on innovations that will improve the Nation's ability to predict, detect, and control wildfires. Recently, DHS announced successful trials of four prototype early detection wildfire sensors. Through a variety of sampling sensors and advanced algorithms, these technologies, once deployed, will provide early warning to fire-vulnerable communities, ultimately providing valuable time that will aid responders and save lives.[136]

Technology exists today that can detect fires within five minutes of when they start. For example, an Australia-based company satellites detected the 2019 Kincade Fire in California wine country 66 seconds after it started, and the fire was verified within three minutes by a network of cameras and sensors coordinated by three universities.

NASA's satellites and high-altitude aircraft can monitor the thermal signatures of fires and gauge the amount of available fuel. The National Interagency Fire Center coordinates drones that collect more infrared fire data and images closer to the ground. California's CALFIRE integrates data from these sources, classified government sources, and the GPS of firefighters on the ground to create simulations for real-time decision-making. Not everyone has easy access to that data, though. Free, accurate data about where and how fires are moving, with a single data clearinghouse, would allow local governments and companies to tailor technology to community-specific needs, such as evacuation planning and better zoning to minimize fire damage.

Investing in more satellite coverage and other sensors would also improve detection and tracking. With a national strategy, the government could invest in research that pulls together many of the agencies involved in research, planning, and response. The Departments of Agriculture and Interior, NASA, the U.S. Geological Survey, the National Oceanic and Atmospheric Administration, the National Institute of Standards and Technology, the National Institutes of Health, the Defense Department, the Space Development Agency, and others are all involved in research that could benefit fire management efforts.[137]

---

136 The White House, *Fact Sheet: The Biden-Harris Administration Acts to Address the Growing Wildfire Threat,* https://www.whitehouse.gov/briefing-room/statements-releases/2021/06/30/fact-sheet-the-biden-harris-administration-acts-to-address-the-growing-wildfire-threat/, June 30, 2021

137 Stavros, Natasha, *The Conversation, Bringing tech innovation to wildfires: 4 recommendations for smarter firefighting as mega-fires menace the US,* https://theconversation.com/bringing-tech-innovation-to-wildfires-4-recommendations-for-smarter-firefight-

There are also possibilities that private enterprises could also provide solutions. In the future, private satellites could detect fires as soon as they start and artificial intelligence software dispatches firefighting drones. This "Silicon Valley" solution featuring artificial intelligence (AI) and flying robots is perhaps unsurprising. There are already fledgling companies starting up. A San Francisco company already has four satellites in orbit that can provide detailed landscape photos day or night, through clouds or smoke, and could provide rapid information to the people on the ground to ensure that, when they go into an area, they know what to expect.[138]

Where fire tower lookouts once provided a necessary function for identifying and fighting new fires, the most recent developments in using cameras shows a modern revival in their use. Now, with the growth of technology, cameras throughout wilderness areas have helped firefighters and communities react quicker and smarter when fire emerges. Cameras can confirm fires quickly, effectively saving time otherwise used sending engines to the reported location or launching aircraft to fly overhead. ALERT Wildfire was established in 2016 as a congregator of cameras in California, Nevada and Oregon. Anyone with Internet access can visit the site leading to a map of numerous cameras. In addition to quickly spotting fires, the cameras are key to decision making when leaders can immediately gauge the resources they will need and when to evacuate specific areas. It also allows homeowners to monitor fires that could potentially threaten their communities. The cameras, many solar-powered, can pan, tilt, zoom, and perform 360-degree views every two minutes. They provide near-infrared night vision views as well. In clear weather, a camera can view as far as 60 miles away, and at night, 120 miles. There is also a time-lapse function available for both firefighters and the public to

---

ing-as-megafires-menace-the-us-162178, June 8, 2021

138 Baron, Ethan, *New Technology Aims to Help Fight California Wildfires*, https://www.govtech.com/news/new-technology-aims-to-help-fight-california-wildfires, Mercury News, June 28, 2021

review. The easily accessible camera network site has been particularly helpful for local fire watch groups. The network has doubled in recent years to more than 700 cameras across six states. The State set a goal of 1,000 cameras by 2023.[139]

ALERT Wildfire is a consortium of three universities — The University of Nevada, Reno (UNR), University of California San Diego (UCSD), and the University of Oregon (UO) — providing access to state-of-the-art Pan-Tilt-Zoom (PTZ) fire cameras and associated tools to help firefighters and first responders: (1) discover/locate/confirm fire ignition, (2) quickly scale fire resources up or down appropriately, (3) monitor fire behavior through containment, (4) during firestorms, help evacuations through enhanced situational awareness, and (5) ensure contained fires are monitored appropriately until extinguished.

A new strategy was adopted by ALERT Wildfire in early 2018 to install cameras on existing third-party microwave networks, build larger virtual networks, produce regional coverage, and do it quickly. In this model, "towers of opportunity" (e.g., Wireless Internet Service Providers, state and county services, and other private point-to-point communications infrastructure) are outfitted with fire cameras and associated equipment to potentially allow one hundred or more fire cameras to be installed in a single season. The data from these confederated networks are seamlessly incorporated into a back-end acquisition systems and presented on a website. To firefighters and first responders, it means "more cameras, more quickly," which equals better decision-making capabilities. Now dozens and dozens of cameras can be installed in a single month with the goal of 1000 cameras in California by 2022; efforts to scale up in other states and countries

139 Prell, Adam, RedZone, *How Cameras Benefit Fire Fighting Efforts*, https://www.redzone.co/2022/04/29/how-cameras-benefit-fire-fighting/?utm_campaign=External%20Blog&utm_medium=email&_hsmi=211623536&_hsenc=p2ANqtz-8nQwkoerHMDI-SgK-d1pGSioW43xzy2OxHknsekVGf93PFj4JfLfIaF1ienpLiMB4qwgQ-V9-677UtoUtbABvY-sWM_B_Q&utm_content=211623536&utm_source=hs_email, April 29, 2022, accessed April 30, 2022

are underway.[140] The path to the future seems clear for this system. Integration with existing systems such as lightning detection would be a likely improvement (Figure 101).

In Michigan, the return to the tower system for fire detection is foreseeable. That future system of fire towers could be augmented with

Figure 101. Technicians installing an ALERT Wildfire camera

artificial intelligence; with inputs from a variety of satellite systems, a wide array of sensors including lightning detection, and cameras looking at the forest, tirelessly 24/7.

DRONES

The drone program, in the MDNR Forest Resources Division, has been expanding since it started with just three vehicles in 2016. By the middle of 2017, the MDNR program had three drones and a certificate of authorization from the Federal Aviation Administration. It's the

---

140 Michell, *Pillar Point, Alert Fire, and Smoke Point Use Harbor Camera Now Pointing at El Granada Hills*, https://www.coastsidebuzz.com/pillar-point-alert-fire-and-smoke-point-use-harbor-camera-now-pointing-at-el-granada-hills/, August 3, 2021, accessed April 30. 2022

second Michigan agency to receive a certificate, after the Michigan State Police. In 2020 the MDNR had seven licensed pilots and 10 drones of different sizes and capabilities. Today, fires are mapped quickly by MDNR drones with verification by firefighters on the ground. They are also using drones to use infrared imaging to find hot spots within the containment line that need attention.[141]

Michigan is also exploring the use of drones to fight wildland fire. Michigan was one of the first Midwestern states to purchase an IGNIS firefighting drone, a tool that is being used with great success nationally.[142] IGNIS is equipped with a hopper that carries a payload of 450 standard ping-pong ball-sized chemical spheres called "dragon eggs" (Figure 102). The spheres are pre-loaded with potassium permanganate and later injected with antifreeze prior to being released. The operation of the drone and IGNIS is tightly controlled to allow the firefighters to release the spheres with a high degree of accuracy. The spheres ignite after hitting the ground, setting off back burns to create defensive lines for firefighters in rugged and sometimes

Figure 102. Drone dropping "dragon eggs"

141 Heckman, Kerry, PIO, Michigan.gov, *Firefighters have Blue Lakes Fire 75% contained, drones deployed to map hot spots,* https://www.michigan.gov/dnr/about/newsroom/releases/2022/05/16/blue-lakes-fire-75-percent-contained, May 16, 2022

142 News-Herald, Michigan Department of Community Health, Michigan Department of Natural Resources, and Michigan Department of Transportation, *A view from above: Michigan DNR explores drone technology,* https://www.thenewsherald.com/2020/06/28/a-view-from-above-michigan-dnr-explores-drone-technology/, Published: June 28, 2020, updated: June 17, 2021, accessed May 1, 2022

unreachable terrain. It can be integrated with infrared and other types of cameras. This gives the drone the dynamic ability to fly in dense smoke, minimizing risk to firefighters on the ground and to high-risk, low-flying manned aircraft.[143]

A heavy-lift drone is now available, weighing in at 120 pounds, that can carry 100 pounds of cargo plus fuel and fly for up to two hours fully loaded. Affectionately likened to an aerial pickup truck by the manufacturer, the Firefly heavy-lift drone is designed to operate as a step-up hauling vehicle from existing equipment-transporting UAVs to helicopters. The craft – whose current top speed is 100 mph – will eventually be developed to carry as much as 1,000 lbs. Heavy-lift drones will become workhorses that can bring supplies to firefighters on the front lines, drop off payloads, and eventually put out small fires.[144]

It is easy to imagine a future where a network of camera towers with satellite and other ground sensor input, coordinated by artificial intelligence and human fire managers, could dispatch drones, and suppress wildfires within a fraction of the time it takes today.

COMMUNICATIONS

We have seen many examples of fast, unpredictable movements and changes in wildfires that have resulted in the death of firefighters. In 1994, the South Canyon Fire in Colorado killed 14 firefighters because of a growing spot fire. In 2013, strong winds overran 19 hotshots in Yarnell Hill, Arizona. All 19 firefighters died because command was unable to locate them. But situations like this don't have to occur. We may not be able to prevent these tragic deaths altogether, but we can lower the risk with better communication. To help reduce the

143 Drone Amplified Editor, U.S. *Forest Service and Drone Amplified partner to drive the search for fire-fighting drones*, https://droneamplified.com/u-s-forest-service-and-drone-amplified-partner-to-drive-search-for-domestic-fire-fighting-drones/?v=7516fd43adaa, September 9, 2020, accessed May 1, 2022

144 Crumley, Bruce, *Firefly (really) heavy-lift drone boasts mega-endurance, too*, https://dronedj.com/2022/01/04/firefly-really-heavy-lift-drone-boasts-mega-endurance-too/, January 4, 2022, accessed May 1, 2022

risk posed by wildfire response, it's essential that everyone remains connected and can easily identify the location of all firefighters in the area. To accomplish this, the ideal communications system would allow crews to track the locations of every individual firefighter, engine, tractor-plow, bulldozer, and aircraft. It would also show their locations on the same screen as the location of the fire. It should also give users the ability to message each other quickly and efficiently, even if voice radio communications are denied. The future is "now" for this system, and it is called a mobile mesh network. It gives firefighters awareness of exactly where the fire is burning and shows senior leaders and commanders where their crews are in relation to the fire so that they can direct them to safety, if necessary.[145]

Mobile mesh networks form by creating peer-to-peer (or device-to-device) connections between mobile phones or IoT (internet of things) devices. Connections form through Bluetooth, Wi-Fi direct, and existing phone technologies. Once a device connects to a mobile mesh network, it can send messages, files, data, etc., in one of two ways. Data can be sent in a single 'hop' directly from one device to another. Or data can be delivered in a 'multi-hop' manner, with messages being transferred through many nodes (devices) until it reaches its endpoint. This method facilitates offline message delivery, over greater distances. Imagine you're at a music festival, in a remote area, with no cell phone service. You came with your friend, but amidst the crowd, you've been separated. Although you have no connectivity bars, you do have a mesh-enabled messaging app. You're able to send a message from one side of the festival that hops from device to device securely, without being made visible to anyone, until it reaches your friend on the other side (Figure 103).[146]

---

145 Schradin, Ryan, *Four considerations for scaling mesh networks on the fireline*, https://thelastmile.gotennapro.com/four-considerations-for-scaling-mesh-networks-for-wildfire-response/, October 19, 2020, accessed May 2, 2022

146 McLennan, Amber, RightMesh, *What is Mobile Mesh Networking?*, https://medium.com/rightmesh/what-is-mobile-mesh-networking-9647328814943, March1, 2019, accessed May 2, 2022

The mobile mesh solution being used on wildfires presently is ATAK, which stands for Android Team Awareness Kit, shortened to Team Awareness Kit (TAK) in some jurisdictions. Created for the military over ten years ago, it currently has a user base that extends well outside of the U.S. Department of Defense (DoD). Today, the application has 40,000 DoD users, as well as 32,000 nonmilitary users.[147] Its use by the State of Colorado is a good example. They have brought together ruggedized smartphones, the ATAK application, and paired goTenna Pro X mobile mesh networking radio devices to create an off-grid situational awareness and communications system for wildland fire response. This solution gives commanders the ability to see their firefighters' locations on a map, see where those locations are in relation to the fire, and even communicate with them via text — even when terrestrial mobile and radio infrastructure is denied

Figure 103. Using a mobile mesh network

or compromised.[148] The network consists of goTenna Pro units that

---

147 Stewart, Luke, *The Last Mile, Situational awareness? Thanks to ATAK, there's an app for that!*, https://thelastmile.gotennapro.com/situational-awareness-thanks-to-atak-theres-an-app-for-that/, February 4, 2021, accessed May 2, 2022

148 Schradin, Ryan, *Four considerations for scaling mesh networks*, May 2, 2022

connect via radio waves and operate independently of wireless internet and traditional cell service. Users connect to the system via their phones and the goTenna Pro app. It pairs with an iOS or Android device through Bluetooth, creating a mesh network that allows for one-to-one or one-to-many messaging and the ability to create a private network. [149]

When using a goTenna Pro X device on a mesh network (Figure 99), firefighting operators can deploy, coordinate, and communicate with fire crews, contract hires, search and rescue teams, and other interagency operations throughout an incident's entirety. Operators are also able to monitor and share critical messages on the latest fire perimeter and hazard updates, in real-time. Providing a clear operating picture of the ground-based fire line enables efficient, organized, and strategic decision-making for firefighters on the ground. If circumstances on the ground were to evolve to a point where responding crew members found themselves in a life-threatening situation, responders with mesh network connectivity can access an application on their mobile devices that has the functionality to map real-time escape routes to safety zones and other points of interest.[150]

Wildland fire response is not just contained to the ground. Another key domain that is leveraged during wildfire response is the air, where helicopters and planes provide critical, operational support through above-ground monitoring and water or retardant drops. Due to the hazardous nature of aerial operations, it is crucial that responders on the ground – and in the air – have complete situational awareness of each other, so that crews in the air can execute drops safely and not interfere with operators below. Through the real-time situational awareness capabilities that mobile mesh network devices provide to

---

149 McCay, Jim, Government Technology, *Mesh Network Creating Situational Awareness During Wildfires*, https://www.govtech.com/em/disaster/mesh-network-creating-situational-awareness-during-wildfires.html, April 5, 2019

150 Pesgraves, David, *The Last Mile, 3 mesh network use cases for wildland fire deployment*, https://thelastmile.gotennapro.com/3-mesh-network-use-cases-for-wildland-fire-deployment/, April 7, 2022, accessed May 2, 2022

wildfire response teams, crews in the air can coordinate and execute their operations effectively and safely, preventing any injuries or loss of life of responders on the ground. For responders operating in helicopters or planes, noise and sound obstructions can often drown out voice communications over a radio. In scenarios when voices cannot be heard, operators in the air using mesh network-enabled devices and applications can still communicate with team members on the ground via text messaging.[151]

State, Federal and local cooperators could benefit significantly from the usage of mobile mesh networking. It seems that the future is already here for this technology.

EVACUATION

In California, during the Santa Cruz Mountains Fire in 2020, one in a series of huge blazes sparked by dry lightning, Bay Area startup Zonehaven's map-based evacuation software for official and citizen use was put to the test. Zonehaven was founded in 2018, and already, Cal Fire and dozens of other agencies and local governments — including Santa Clara, San Mateo, Contra Costa, and Alameda counties — are adopting it to coordinate the safe exodus of people from threatened areas. The public app shows residents where they are on a map, with evacuation status — from advisory to warning to order — shown by the coloring of their zone. In the era of mega-fires, evacuation of the public will become a top priority, and planning for that is an immediate need.[152]

COMPUTER MODELING

Once ignited, wildfires can spread at varying speeds and change direction in a matter of seconds, making blazes tricky to predict.

---

151 Pesgraves, David, *The Last Mile, 3 mesh network use cases,* accessed May 2, 2022

152 Baron, Ethan, *New Technology Aims to Help Fight California Wildfires*, https://www.govtech.com/news/new-technology-aims-to-help-fight-california-wildfires, Mercury News, June 28, 2021, accessed May 3, 2022

Most agencies do it manually, looking at the weather, terrain, and the dryness of the vegetation. But coming up with calculations can be time-consuming, an eternity when facing a fast-spreading fire. Fire agencies in California are getting some help from a powerful new tool. FireMap (https://firemap.sdsc.edu/), an artificial intelligence-based platform developed by WIFIRE Lab, a spin-off of the San Diego Supercomputer Center (SDSC) at the University of California, San Diego, can create in minutes, a predictive map of the fire's expected trajectory. The system builds on a combination of deep learning techniques to crunch real-time data about weather, topography, the dryness of vegetation, and more from satellites, on-the-ground sensors, utility cameras, and, more recently, a fixed-wing aircraft outfitted with infrared radar. This information is brought together and fed into models that can predict where the fire will be, its rate of spread, and its direction for up to six hours. Those predictions help incident commanders make critical judgment calls, such as where to send their limited firefighting personnel and whether to issue evacuation orders. Today, LAFD and several other fire departments across Southern California routinely put the fire-predicting system to work for fighting the area's increasingly dangerous wildfires; about 130 other groups are testing the technology.[153]

A NEW FRAMEWORK

In a report to the President, released in February 2023, the President's Council of Advisors on Science and Technology (PCAST) issued a framework to guide the implementation of emerging technology into wildfire control.[154] To do so, PCAST sought extensive input from

---

153 Rossi, Marcello, *From Supercomputers to Fire-Starting Drones, These Tools Help Fight Wildfires*, Smithsonian Magazine, https://www.smithsonianmag.com/innovation/from-supercomputers-to-fire-starting-drones-these-tools-help-fight-wildfires-180978421/, August 20, 2021, accessed May 3, 2022

154 President's Council of Advisors on Science and Technology, *Report to the President, Modernizing Wildland Firefighting to Protect Our Firefighters, Executive Office of the President*, https://www.whitehouse.gov/wp-content/uploads/2023/02/

more than one hundred experts and stakeholders, including science, technology, military, and wildland firefighting.

The report "highlights immediate needs that can be addressed with existing technology, and strategic, long-term investments in new science and technology to ensure that our firefighters do not have to face tomorrow's fires with yesterday's tools." PCAST's report states, "Whereas we have a national commitment ensuring that our warfighters are not sent into harm's way without the best of American science and technology at their disposal, no similar organizational framework exists to protect and empower wildland firefighters. We recommend you establish a new joint executive office with Cabinet-delegated authorities to implement a unified science and technology strategy for wildfires...."

Five main recommendations were developed by PCAST to better integrate emerging technologies into government firefighting operations amid the continued prevalence of large-scale wildfires. They first recommended the immediate assessment, adaptation, and field currently available communications technologies, such as mobile area networks, that are now commonly used in the commercial and defense sectors. These networks can enhance communications in terrain that is incredibly challenging and dangerous for wildfire response.

Second, they recommended the establishment of a joint-agency executive office featuring Cabinet-delegated authorities to formally develop and implement a science and technology strategy tailored for effective wildfire fighting. It would serve to advance coordination, streamline authorities, and drive progress in enabling technology adoption across the numerous federal agencies

Third, the new Joint Office would strengthen the whole operational sequence of wildland firefighting— detection, alert, response, and suppression—by assessing existing technologies available within the federal arena, the private sector, and allied nations. It would incorporate

PCAST_Wildfires-Report_Feb2023.pdf, February, 2023, Accessed March 1, 2023 other technologies like uncrewed aerial and ground vehicles, drones and robotics, commercial satellite feeds, and field sensors into modern wildfire management. These technologies would be autonomous and ideally analyze and process data to detect and suppress potentially catastrophic wildfire situations.

Fourth, they recommended improving predictive wildfire modeling tools by engaging a broader community of researchers in national labs, universities, and the private sector—with expertise in artificial intelligence, atmospheric science, remote sensing, advanced computational methods, and other disciplines central to the understanding of wildfires. Predictive wildfire modeling tools would be improved by expanding research community access to archived satellite data from defense and other government sources. Effective artificial intelligence modeling of wildfire spread based on terrain, vegetation cover, soil moisture, wind, and other factors with much broader access to abundant historical data—some of which exists in defense archives but is currently classified.

Fifth, PCAST recommended expanding our nation's wildfire response capacity by encouraging the development of prototype autonomous detection, assessment, and containment systems for wildland fire. Uncrewed aerial vehicles and other autonomous systems would dramatically increase our nation's wildfire response capacity, especially at a fire's developing stages, while providing new means to protect firefighters from active wildfires.

Presidential action on these recommendations could address growing shortages of firefighting resources. It would also leverage our nation's leading science and technology to protect wildland firefighters better and modernize our wildfire response enterprise, including the tools used to combat fires and the organizations that develop and deploy them.

THE FUTURE OF WILDFIRE IN THE UPPER PENINSULA

The future of wildfire in the Upper Peninsula and the Seney NWR shows rising temperatures and drier weather by the end of the century, according to recent studies, and that fire probability will increase and could double in the Great Lakes region. Warming will likely increase the potential for 'very large fires'—the top ten percent of fires, which account for a majority of burned areas in many regions of the United States. Precipitation increases in the fall and spring may contribute to vegetation growth that would dry out and die during hotter, long summer days without rain and with increased chances of lightning, providing the setting for more severe wildfires. These environmental changes, combined with an increased human presence, wildfires will be a more significant threat in the future under a changing climate.[155]

The increase in the number of "very large fire weeks"—weeks in which conditions are favorable to the occurrence of very large fires—by mid-century (2041-2070) compared to the recent past (1971-2000) is expected to increase in the forests around the Great Lakes.[156] Scientists predict the total area burned will increase between 25% and 53% by 2100.[157]

There are several variables when considering the future of wildfires in the Seney NWR. If climate change continues its present arc, the upper Midwest and the Upper Peninsula will have longer, warmer, and drier fire seasons. The potential for larger fires will increase with more extended periods of fire weather conducive to burning. There will be an increase in human-caused fires, partly due to climate migration, and a greater possibility of lightning-caused fires. Based

155 Neumann, Brad, Climate Change and Wildfire in the Great Lakes Region, Michigan State University, Extension Bulletin E-3277, August 2015

156 Kennedy, Caitlyn, Risk of very large fires could increase sixfold by mid-century in the US, Climate. Gov, https://www.climate.gov/news-features/featured-images/risk-very-large-fires-could-increase-sixfold-mid-century-us, published August 26, 2015, updated July 2, 2021, accessed April 11, 2022

157 Amadeo, Kimberly, How Wildfires Impact the Economy, https://www.thebalance.com/wildfires-economic-impact-4160764, accessed June 11, 2021

on fire history, it is likely that lightning will again cause fires in the most remote areas of the Seney NWR, particularly during extreme drought conditions. In the future, fires that start in extreme conditions will be professionally managed by the Refuge staff and suppressed by well-trained, qualified firefighters. They will use the latest equipment and technology, including autonomous detection, assessment, and containment systems such as uncrewed aerial vehicles and other autonomous systems. The increased use of prescribed fire could mitigate larger fires through fuel reduction, but some will escape and become large problematic fires.

Firefighting agencies in Michigan, and the Upper Peninsula, will need to move from short-term fixes to transformational change to address the long-term picture; this will be challenging. Barring a significant difference in present climate trends, increased cooperation at all levels will become necessary. Wildfire preparedness will require local, State, and Federal agencies to foster and support a partnership culture before rather than during a disaster; this includes understanding their authorized roles and responsibilities during more complex, multi-agency wildfire emergencies. As wildfire events become more complex and new decision support tools become available, enhanced training and education will be required to strengthen wildfire management capabilities.

Wildfires will continue to be problematic well into the foreseeable future, and determining how best to work together to manage them safely and effectively will be a monumental undertaking. Increased preparedness capacity will be required for the future by enhancing firefighting capacity, including training; renewing critical assets; maintaining or replacing essential infrastructure, and developing innovative solutions to overcome human resource challenges. Agencies must better leverage the nation's existing and emerging science and technology to protect wildland firefighters better and modernize our wildfire response enterprise, including the tools used to combat fires and the organizations that develop and deploy them.

APPENDIX

WEATHER DATA: PAGES 189-196

Weather observations are collected from a standard observation site and time with observations to be taken at solar noon when the sun is at its peak directly overhead. In 1976, weather data taken at the MDNR Seney Field Office was collected manually once a day at 1:00 p.m. EST from their standard observation site. Weather record abbreviations are: Temp=Temperature (F); RH=Relative Humidity(%); Wind=Windspeed in miles per hour; Wind Dir=Direction the wind is coming from; Dur Hrs=Hours of precipitation; Rn24=Amount of precipitation during the past 24 hours. Fire danger rating, R, refers to the daily fire danger rating computed from the weather data gathered. Adjective ratings are: L=Low, M=Moderate, H=High, VH=Very High, and E=Extreme. Fire Weather Index System readings calculated in the CFFDRS are: FFMC=Fine Fuel Moisture Code; DC=Drought Code; ISI=Initial Spread Index; BUI=Buildup Index; and KBDI=Keetch-Byram Drought Index. Weather data and related CFFDRS codes and indexes on the following pages are for the period April 12 through October 31, 1976. The data was provided courtesy of the Michigan DNR.

PALMER DROUGHT SEVERITY INDEXES: PAGES 197-200

Monthly maps of drought conditions in the contiguous U.S. in 1976 as measured by the Palmer Drought Severity Index were copied from: https://www.ncdc.noaa.gov/temp-and-precip/drought/historical-palmers/maps/psi/197605-197704

FIRE BEHAVIOR TABLES: PAGES 201-202

Fire behavior was estimated for the Diary of the Fire utilizing two Canadian Forest Fire Danger Rating System (CFFDRS) indices, the Buildup Index (BUI) and Initial Spread Index (ISI). The BUI combines the current Duff Moisture Code (DMC) and Drought Code (DC) to produce an estimate of potential heat release in heavier fuels. The Initial Spread Index (ISI) integrates fuel moisture for fine dead fuels and surface windspeed to estimate a spread potential. The ISI is a key input for fire behavior predictions.

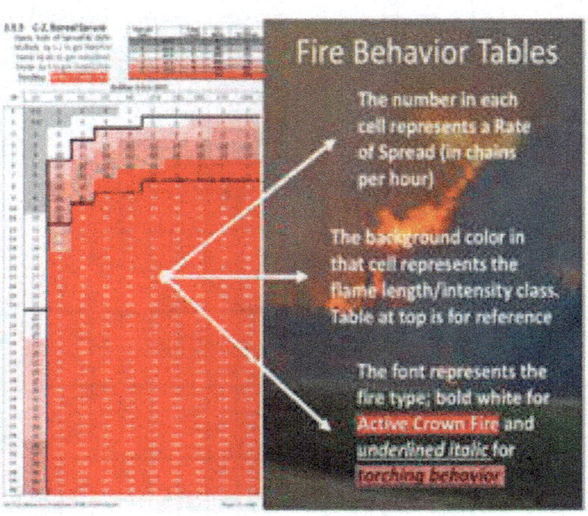

Figure 104. Fire Behavior Table usage

Using weather data from the Seney Field Office, daily outputs of the CFFDRS and its Fire Behavior Prediction (FBP) system, estimates of Rate of Spread (ROS) and Flame Length (FL) were made for each day in the daily fire diary. Two fuel models from the Michigan Field Guide for CFFDRS Fire Behavior Prediction (FBP) System were used; C-2 (Boreal Spruce) and O-1b 70% cured (Standing Grass). Rate of Spread (ROS) is in chains per hour (a chain is 66 feet long) and flame length is in feet. Rate of Spread is given for a fire running with the wind only (Figure 104).

MDNR PARTIAL PERSONNEL LIST: PAGES 203-210

PERIMETER MAP SHOWING ORIGIN: PAGE 211

Weather data, CFFDRS codes and indexes
April 12 to May 13, 1976
Michigan DNR Field Office-Seney, MI

Date	Temp	RH	Wind	Wspd	Dur	Rn24	FFMC	DMC	DC	ISI	BUI	FWI	DSR	R	KBDII
	°F	%	Dir	mph	hrs	inch									
4/12/76	44	50	SW	10	0	0	85.2	6.9	17.2	4.8	6.9	4.3	0.4	High	0
4/13/76	62	25	SW	11	0	0	90.4	10.1	21.2	11	10	11	1.9	VeryHigh	0.2
4/14/76	53	64	SE	13	1	0.11	69.6	8.3	24.3	1.8	9	1.1	0	Low	2.5
4/15/76	64	70	S	3	0	0	77.3	9.7	28.5	1.1	10.5	0.7	0	Moderate	4.8
4/16/76	67	58	SE	10	1	0.01	83.8	11.8	33	4	12.4	4.9	0.5	High	8.2
4/17/76	59	89	E	16	6	0.16	52.2	7.9	32.2	0.8	9.8	0.5	0	Low	11.6
4/18/76	57	60		13	0	0	73.5	9.4	35.7	2	11.3	2	0.1	Moderate	16.3
4/19/76	56	45	N	10	4	0.3	59.5	6.6	28.3	0.9	8.3	0.5	0	Low	8.7
4/20/76	59	39	N	4	0	0	78.3	9	32	1.3	10.5	0.8	0	Moderate	10
4/21/76	50	87	E	6	3	0.29	36	4.8	24.6	0	6.5	0	0	Low	2.6
4/22/76	39	92	N	12	8	0.13	28.4	2.8	23.2	0	4.3	0	0	Low	3.1
4/23/76	52	52	N	9	3	0.18	47.9	2.4	20.9	0.3	3.8	0.1	0	Low	3.6
4/24/76	50	33	N	9	0	0	73.6	4.2	23.7	1.5	5.8	0.7	0	Low	4.4
4/25/76	45	39	N	13	0	0	82.6	5.4	26	4.4	7.1	3.9	0.3	High	5.1
4/26/76	33	90	NW	16	0	0	79.7	5.4	27.1	4	7.2	3.6	0.3	High	5.6
4/27/76	42	62	NW	10	0	0	81.6	6	29.1	3	7.9	2.7	0.2	Moderate	5.6
4/28/76	50	33	NW	12	0	0	86.8	7.8	31.9	7.1	9.7	7.4	0.9	High	5.4
4/29/76	59	48	W	5	0	0	86.9	9.8	35.6	4.1	11.6	4.8	0.4	High	5.9
4/30/76	54	65	W	7	0	0	85.8	10.9	38.8	4.1	12.8	5.2	0.5	High	7.7
5/1/76	50	74	SE	10	1	0.03	80.7	11.6	41.6	2.7	13.7	3.4	0.2	Moderate	9.4
5/2/76	48	42	NW	5	4	0.35	51.3	7.2	32.5	0.3	9.3	0.2	0	Low	1.4
5/3/76	35	73	NW	12	1	0.01	63.6	7.4	35.2	1.3	9.7	0.8	0	Low	1.7
5/4/76	55	34	SW	15	0	0	82.5	9.8	39.9	5.1	12.2	6.1	0.7	High	1.1
5/5/76	34	81	NW	8	6	0.15	48.6	6.2	38.4	0.3	8.9	0.1	0	Low	2.5
5/6/76	50	33	SE	6	0	0	72.6	8.2	42.6	1.1	11	0.7	0	Low	3.7
5/7/76	48	60	N	3	0	0	78.3	9.3	46.6	1.2	12.4	0.8	0	Moderate	4.5
5/8/76	61	28	SW	7	0	0	87.8	12.6	51.9	5.5	15.7	7.5	1	High	4.8
5/9/76	68	32	W	4	0	0	90.1	16.4	57.9	6	19.2	9	1.3	High	6.9
5/10/76	60	40	W	3	0	0	90	19	63.1	5.5	21.7	9	1.3	High	10.5
5/11/76	44	57	NW	10	0	0	87.6	19.9	66.7	6.8	22.8	11.1	1.9	High	14.1
5/12/76	59	25	NW	6	0	0	90.4	23.1	71.8	7.4	25.6	12.5	2.4	High	17.7
5/13/76	67	34	S	12	0	0	90.4	26.7	77.7	12	28.7	19.3	5.1	VeryHigh	19.4

Weather data, CFFDRS codes and indexes
May 14 to June 14, 1976
Michigan DNR Field Office-Seney, MI

Date	Temp	RH	Wind	Wspd	Dur	Rn24	FFMC	DMC	DC	ISI	BUI	FWI	DSR	R	KBDII
	°F	%	Dir	mph	hrs	inch									
5/14/76	63	51	N	8	6	0.3	63	16.8	71.3	0.9	21.1	0.9	0	Low	12.4
5/15/76	68	43	S	15	0	0	83.9	20	77.3	6.1	24.3	10.4	1.7	High	14.9
5/16/76	50	74	N	10	11	1.01	38.8	8.6	36.3	0.1	10.8	0	0	Low	3.7
5/17/76	53	58	N	19	0	0	68.4	10	40.8	2.8	12.4	3.3	0.2	Moderate	7.9
5/18/76	54	54	N	15	0	0	80.3	11.6	45.4	3.9	14.2	5.2	0.5	Moderate	8.8
5/19/76	65	17	SW	4	0	0	90.7	15.8	51.1	6.5	17.9	9.4	1.4	High	9.8
5/20/76	66	41	NW	12	0	0	90.5	18.9	56.9	12.1	20.7	16.5	3.9	VeryHigh	12.9
5/21/76	53	42	N	18	0	0	89.6	20.8	61.4	17.4	22.6	22.4	6.7	VeryHigh	16
5/22/76	55	39	NW	18	0	0	89.6	23	66.1	17.4	24.6	23.4	7.2	VeryHigh	16.9
5/23/76	55	44	NW	15	0	0	89.1	25	70.8	12.7	26.6	19.3	5.1	VeryHigh	18
5/24/76	53	53	N	15	0	0	87.9	26.6	75.3	10.6	28.2	17.4	4.3	VeryHigh	19.1
5/25/76	62	42	NE	6	0	0	88.2	29.3	80.7	5.3	30.7	10.6	1.8	High	20.2
5/26/76	69	44	S	10	0	0	88.5	32.5	86.8	7.7	33.6	15	3.3	High	22.6
5/27/76	72	39	S	12	0	0	89.6	36.2	93.2	10.6	36.8	19.8	5.4	VeryHigh	26.4
5/28/76	76	35	SE	6	0	0	90.7	40.6	100	7.6	40.6	16.4	3.8	High	32.4
5/29/76	70	56	E	5	0	0	88.9	43.2	106.2	5.5	43.2	13.1	2.6	High	38.7
5/30/76	61	60	NE	6	6	0.35	55.3	24.7	96.1	0.5	30	0.6	0	Low	29.4
5/31/76	66	60	E	5	4	0.29	52.8	15.6	90	0.3	21.8	0.3	0	Low	25.8
6/1/76	72	25	N	6	0	0	84.2	20.2	96.4	3.1	26.5	6	0.6	Low	28.9
6/2/76	73	26	N	7	0	0	91.2	24.9	103.9	8.9	31.1	16	3.7	High	33.5
6/3/76	79	32	S	5	0	0	91.7	29.8	112	8.1	35.8	16.1	3.7	High	38.7
6/4/76	86	29	NE	7	0	0	92.9	35.6	120.8	11.4	41	22.1	6.5	High	47.6
6/5/76	87	27	S	7	0	0	93.5	41.7	129.7	12.3	46.2	24.8	8	VeryHigh	59.3
6/6/76	88	26	SW	5	0	0	93.8	48	138.7	11	51.5	24.1	7.6	VeryHigh	70.8
6/7/76	85	28	W	8	0	0	93.9	53.8	147.4	14	56.3	29.8	11.1	VeryHigh	82.1
6/8/76	80	48	SW	3	1	0.02	90.6	57.6	155.6	5.9	59.8	16.7	4	Moderate	94.5
6/9/76	78	57	W	6	0	0	89.1	60.6	163.6	6.1	62.9	17.6	4.3	High	106.7
6/10/76	80	43	SW	5	0	0	89.7	64.8	171.8	6.1	66.7	18	4.5	High	114.4
6/11/76	66	49	NW	12	0	0	89.2	67.5	178.6	10	69.4	26.3	8.9	VeryHigh	122.4
6/12/76	69	47	SE	11	0	0	89.2	70.5	185.7	9.3	72.4	25.6	8.4	VeryHigh	130.3
6/13/76	80	45	S	10	0	0	89.4	74.5	193.9	8.7	76	25.1	8.1	VeryHigh	134.1
6/14/76	74	66	S	10	0	0	87.3	76.7	201.5	6.5	78.6	20.8	5.8	High	145.6

Weather data, CFFDRS codes and indexes
June 15 to July 16, 1976
Michigan DNR Field Office-Seney, MI

Date	Temp	RH	Wind	Wspd	Dur	Rn24	FFMC	DMC	DC	ISI	BUI	FWI	DSR	R	KBDII
	°F	%	Dir	mph	hrs	inch									
6/15/76	63	95	S	25	4	0.85	28	33.9	157.1	0	44	0	0	Low	90.9
6/16/76	53	76	NW	15	0	0	54.4	34.7	162.6	0.9	45.3	2.2	0.1	Low	96.1
6/17/76	67	50	SE	10	0	0	78.9	37.4	169.5	2.3	48.2	6.6	0.8	Low	100.6
6/18/76	70	72	S	15	0	0	82.5	39	176.7	5.1	50.3	13.5	2.7	Moderate	104.5
6/19/76	68	39	W	17	5	0.37	72.7	23.3	164.2	2.7	34.4	6.3	0.7	Low	90.4
6/20/76	80	30	NW	11	0	0	90.1	28.4	172.4	10.6	40.2	20.8	5.8	High	93.6
6/21/76	82	26	S	6	0	0	92.9	34	180.8	10.5	46.3	22.2	6.6	VeryHigh	101.4
6/22/76	82	31	SW	5	0	0	93	39.2	189.2	9.7	51.7	22.1	6.5	High	111.1
6/23/76	73	55	S	10	0	0	89.7	42	196.7	9.2	54.8	21.9	6.4	High	118.7
6/24/76	69	68	SW	5	1	0.01	87	43.8	203.8	4.1	57	12.3	2.3	Moderate	126.2
6/25/76	73	55	S	10	1	0.02	86.5	46.6	211.3	5.8	60.1	16.5	3.9	Moderate	131.4
6/26/76	72	46	W	9	1	0.05	84.7	49.9	218.7	4.2	63.6	13.2	2.6	Moderate	137.2
6/27/76	69	59	NW	11	0	0	85.7	52.2	225.8	5.6	66.2	16.9	4.1	Moderate	142.3
6/28/76	70	64	S	12	0	0	85.7	54.3	233	6.1	68.6	18.5	4.7	High	146.8
6/29/76	69	68	E	6	1	0.15	65.5	41.3	233.5	0.9	57.3	2.7	0.2	Low	152.2
6/30/76	69	47	NW	13	0	0	83.6	44.3	240.6	5	60.7	14.7	3.2	Moderate	156.9
7/1/76	68	39	NW	11	0	0	88.3	47.7	247.6	8.2	64.4	22	6.4	High	161.3
7/2/76	70	41	NW	11	0	0	89	50.8	255.1	9.1	67.8	24.2	7.7	High	165.7
7/3/76	80	35	NW	8	0	0	91	55	263.6	9.4	72.3	25.7	8.5	VeryHigh	170
7/4/76	82	31	NE	8	0	0	92.1	59.7	272.3	11	77.1	29.6	10.9	VeryHigh	178.7
7/5/76	87	30	S	7	0	0	92.9	64.9	281.5	11.3	82.3	31	11.9	VeryHigh	189.5
7/6/76	75	51	SE	7	0	0	90.4	67.8	289.5	7.9	85.5	24.9	8.1	High	202.6
7/7/76	75	51	NW	7	5	0.6	65.4	33.9	253.8	0.9	50.8	2.7	0.2	Low	173.3
7/8/76	76	49	S	6	0	0	83.4	37	261.9	2.7	54.6	8.5	1.2	Low	181.1
7/9/76	70	64	S	7	0	0	84.7	38.9	269.4	3.5	57.1	10.9	1.9	Moderate	190.7
7/10/76	80	61	S	8	5	0.5	65.9	20.3	243.6	1	33.6	2	0.1	Low	165
7/11/76	65	61	NW	8	0	0	79.2	22.1	250.6	2	36.2	4.8	0.4	Low	176
7/12/76	64	35	N	9	1	0.01	87.5	25	257.5	6.1	40.2	13.8	2.8	Moderate	186.8
7/13/76	72	32	NE	2	0	0	90.3	28.7	265.2	5.2	45.2	12.9	2.5	Moderate	189
7/14/76	75	66	S	5	7	0.8	54.8	13	217.7	0.4	22.7	0.4	0	Low	134.6
7/15/76	85	45	SE	6	0	0	84.3	16.9	226.7	3.1	28.6	6.3	0.7	Low	142.9
7/16/76	60	58	NW	11	3	0.3	60.2	10.6	216	1	19	0.9	0	Low	141.3

Weather data, CFFDRS codes and indexes
July 17 to August 17
Michigan DNR Field Office-Seney, MI

Date	Temp	RH	Wind	Wspd	Dur	Rn24	FFMC	DMC	DC	ISI	BUI	FWI	DSR	R	KBDII
	°F	%	Dir	mph	hrs	inch									
7/17/76	70	41	NW	12	0	0	83.7	13.7	223.5	4.6	23.7	8.2	1.1	Moderate	143.2
7/18/76	80	41	W	8	0	0	89.3	17.5	232	7.3	29.5	13.4	2.7	Moderate	147.1
7/19/76	85	42	SW	12	0	0	90.3	21.7	241	11.8	35.4	21	6	High	154.3
7/20/76	70	77	NW	9	4	0.31	55.8	12.8	230	0.6	22.4	0.6	0	Low	152.1
7/21/76	75	22	N	5	0	0	86.3	17.4	238	3.8	29.4	7.7	1	Low	156
7/22/76	82	34	S	9	0	0	91.1	21.9	246.7	10.3	35.8	19.2	5.1	High	162.3
7/23/76	72	73	W	10	0	0	86.5	23.4	254.4	5.8	38	12.8	2.5	Moderate	169.4
7/24/76	70	34	N	10	0	0	89.9	26.8	261.9	9.4	42.7	19.6	5.3	High	176
7/25/76	68	51	S	5	0	0	89.2	29.2	269.2	5.7	46	14	2.9	Moderate	180.3
7/26/76	76	56	NW	5	1	0.02	87	31.8	277.3	4.2	49.5	11.4	2	Moderate	185.7
7/27/76	74	51	NW	5	0	0	87.6	34.6	285.2	4.5	53.1	12.7	2.4	Moderate	190.5
7/28/76	63	89	NE	3	1	0.01	82.2	35.1	292	1.9	53.9	5.9	0.6	Low	196.5
7/29/76	78	43	NW	3	0	0	88	38.7	300.3	4.1	58.5	12.4	2.3	Moderate	201.5
7/30/76	79	54	SE	7	1	0.18	74.4	29	298.3	1.3	46.7	3.7	0.3	Low	207.4
7/31/76	69	36	N	11	0	0	87.3	32.3	305.7	7	51	17.4	4.3	High	213.2
8/1/76	70	34	NW	12	0	0	90	35.7	313.2	11.3	55.6	25.6	8.5	VeryHigh	219.7
8/2/76	80	30	NW	9	0	0	92	39.7	321	11.8	60.7	27.5	9.6	VeryHigh	222.7
8/3/76	82	26	SW	8	0	0	93.2	44.1	329	12.8	66.1	30.3	11.4	VeryHigh	229.1
8/4/76	78	43	SW	13	0	0	91.7	47.2	336.6	15.5	69.9	35.4	15	VeryHigh	235.4
8/5/76	67	67	W	9	8	0.25	62.1	29.9	325.2	1	48.6	2.6	0.1	Low	235.7
8/6/76	70	34	NE	3	0	0	82.9	32.9	332	2	52.8	6.3	0.7	Low	238.3
8/7/76	78	28	NW	3	0	0	90.8	36.9	339.6	6.1	58	16.8	4	High	241.7
8/8/76	80	32	NE	6	0	0	91.7	40.8	347.4	8.9	63.1	23	7	High	246.6
8/9/76	83	38	S	7	0	0	91.7	44.6	355.5	9.6	67.9	25.3	8.3	High	252.4
8/10/76	80	45	S	14	0	0	91.1	47.8	363.3	15.6	71.9	36	15.4	VeryHigh	260.9
8/11/76	82	59	SW	11	0	0	89.1	50.2	371.3	9.2	75.1	25.8	8.6	VeryHigh	269.3
8/12/76	80	61	S	5	2	0.25	67.8	32.6	359.7	0.9	53.1	2.5	0.1	Low	270.6
8/13/76	66	37	NW	6	1	0.02	83.6	35.2	366.1	2.8	56.8	8.9	1.3	Low	277.2
8/14/76	66	37	NE	9	0	0	88.4	37.8	372.5	7	60.3	18.9	4.9	High	282.3
8/15/76	67	34	N	8	0	0	89.9	40.6	379	8	64	21.6	6.2	High	284.3
8/16/76	75	35	W	7	0	0	90.7	44	386.3	8.3	68.4	22.8	6.9	High	286.5
8/17/76	80	35	SW	5	0	0	91.2	47.7	394.1	7.6	73.3	22.2	6.6	High	290.5

Weather data, CFFDRS codes and indexes
August 18 to September 18, 1976
Michigan DNR Field Office-Seney, MI

Date	Temp	RH	Wind	Wspd	Dur	Rn24	FFMC	DMC	DC	ISI	BUI	FWI	DSR	R	KBDII
	°F	%	Dir	mph	hrs	inch									
8/18/76	76	56	S	13	0	0	89.2	50	401.5	10.9	76.3	29.3	10.7	VeryHigh	296.1
8/19/76	85	54	S	16	0	0	89.2	52.9	409.8	14	80	35.3	15	VeryHigh	300
8/20/76	85	42	S	14	0	0	90.3	56.6	418.1	13.8	84.5	36.1	15.5	VeryHigh	308.7
8/21/76	81	42	S	12	0	0	90.3	60	426	11.8	88.7	33.3	13.5	VeryHigh	319.3
8/22/76	82	28	N	8	0	0	92.6	64.3	434	11.8	93.8	34.1	14.1	VeryHigh	327.2
8/23/76	75	48	NE	5	0	0	90.8	67	441.3	7.1	97.1	24.6	7.9	High	333.8
8/24/76	80	38	S	8	0	0	90.8	70.6	449.1	9.2	101.3	30	11.2	VeryHigh	339.6
8/25/76	84	35	SE	14	0	0	91.6	74.6	457.3	16.6	106	45	22.9	VeryHigh	345
8/26/76	80	51	SE	9	0	0	90.3	77.4	465.1	9.2	109.3	31	11.9	VeryHigh	350.6
8/27/76	84	56	S	20	0	0	89.5	80.1	473.3	20.2	112.6	52.2	29.8	VeryHigh	360.8
8/28/76	73	30	S	11	2	0.06	87.8	81	480.4	7.6	114	27.7	9.7	High	366.2
8/29/76	52	41	W	18	0	0	87.8	82.5	485.4	13.5	115.8	40.8	19.3	VeryHigh	372.2
8/30/76	66	41	SE	9	0	0	88.6	84.9	491.8	7.2	118.6	27.3	9.5	High	374.9
8/31/76	64	32	NW	7	0	0	90	87.6	498	7.5	121.6	28.2	10.1	High	376.6
9/1/76	53	48	W	12	1	0.1	74.2	72	503.1	1.9	106.1	9.5	1.5	Low	378.5
9/2/76	62	42	SE	10	0	0	84.8	73.8	507.8	4.6	108.3	19	5	High	379.8
9/3/76	70	41	SE	20	0	0	88.6	76.1	513.3	17.7	111.1	47.8	25.6	VeryHigh	381
9/4/76	59	63	W	9	1	0.04	81.7	77.2	517.7	2.8	112.4	13.4	2.7	Moderate	383.5
9/5/76	62	37	NW	8	0	0	87.4	79.2	522.4	5.6	114.9	22.5	6.7	High	386.7
9/6/76	72	42	SE	16	0	0	89	81.6	528.1	13.4	117.7	41	19.5	VeryHigh	388.3
9/7/76	81	45	S	7	0	0	89.4	84.4	534.7	6.9	121	26.6	9.1	High	390.9
9/8/76	89	31	SE	12	0	0	92.7	88.4	542.1	16.6	125.6	48	25.7	Extreme	396
9/9/76	60	68	W	12	1	0.1	72.3	72.1	546.6	1.8	108.5	9	1.3	Low	403
9/10/76	65	43	W	15	0	0	85.2	74.1	551.6	7.3	110.9	26.7	9.1	High	409.9
9/11/76	75	41	W	7	0	0	88.9	76.7	557.6	6.4	114.2	24.7	7.9	High	411.3
9/12/76	84	27	SW	12	0	0	92.9	80.6	564.5	17.1	118.8	48.1	25.8	Extreme	417.3
9/13/76	78	43	SE	19	0	0	91.5	83.3	570.8	24.7	122.1	61	39.3	VeryHigh	422
9/14/76	54	88	W	10	7	0.38	38.1	44.4	522.9	0.1	73.3	0.1	0	Low	408.3
9/15/76	54	76	NE	5	0	0	55.9	45	526.8	0.5	74.1	1.1	0	Low	411.7
9/16/76	67	50	N	5	0	0	77.5	46.8	532	1.3	76.8	5.6	0.6	Low	412.2
9/17/76	72	35	NE	3	0	0	87.5	49.5	537.7	3.8	80.5	14	2.9	Moderate	414
9/18/76	72	57	W	4	0	0	87.5	51.3	543.4	4.1	83	15.3	3.4	High	416.3

Weather data, CFFDRS codes and indexes
September 19 to October 20, 1976
Michigan DNR Field Office-Seney, MI

Date	Temp	RH	Wind	Wspd	Dur	Rn24	FFMC	DMC	DC	ISI	BUI	FWI	DSR	R	KBDII
	°F	%	Dir	mph	hrs	inch									
9/19/76	64	75	N	10	3	0.19	60	35.1	527.9	0.9	60.2	3	0.2	Low	419.7
9/20/76	56	40	W	8	0	0	79	36.6	532	2	62.5	6.9	0.8	Low	423.4
9/21/76	42	41	W	7	0	0	83.5	37.3	534.7	3	63.5	10.2	1.6	Moderate	425.8
9/22/76	46	27	SW	12	0	0	87.9	38.5	537.8	8.3	65.2	22.3	6.6	High	431.6
9/23/76	46	52	W	22	7	0.14	68.4	29.3	528.4	3.5	51.5	10.2	1.7	Moderate	436.2
9/24/76	52	63	SE	0	0	0	73.8	30.1	532.1	0.7	52.7	1.9	0.1	Low	436.4
9/25/76	54	38	S	3	0	0	82.8	31.6	536	2	55	6.4	0.7	Low	437.1
9/26/76	56	61	SE	3	0	0	83.8	32.6	540.1	2.3	56.7	7.3	0.9	Low	438.1
9/27/76	49	67	N	7	0	0	83.8	33.2	543.5	3.2	57.6	9.9	1.6	Moderate	438.7
9/28/76	54	59	S	13	0	0	84.4	34.2	547.4	5.5	59.1	15.7	3.6	Moderate	439.4
9/29/76	61	54	N	7	0	0	85.6	35.6	552	4	61.3	12.5	2.4	Moderate	440
9/30/76	65	53	SW	12	0	0	86.4	37.2	557	6.7	63.8	19	5	High	441
10/1/76	76	39	SW	11	0	0	89.6	40	563.1	9.9	67.9	25.7	8.5	High	442.3
10/2/76	67	54	N	7	0	0	88.6	41.4	567.3	6.2	70.1	18.7	4.9	High	445.1
10/3/76	70	30	SE	11	0	0	90.9	43.8	571.8	11.9	73.5	30.3	11.4	VeryHigh	448.2
10/4/76	68	55	S	13	0	0	88.9	45.2	576.1	10.4	75.6	28.2	10	VeryHigh	450
10/5/76	48	86	W	1	16	1.12	21.8	17.1	425.9	0	31.1	0	0	Low	360
10/6/76	41	84	N	8	1	0.02	36.4	17.2	427.5	0	31.3	0	0	Low	362.6
10/7/76	44	50	N	2	0	0	56.3	17.8	429.4	0.4	32.2	0.5	0	Low	363.9
10/8/76	48	54	S	4	0	0	70.9	18.5	431.7	0.9	33.4	1.5	0.1	Low	363.9
10/9/76	48	54	NE	3	0	0	78.2	19.2	434	1.2	34.6	2.6	0.1	Low	364.2
10/10/76	50	50	SE	6	0	0	82.9	20	436.5	2.6	36	6.1	0.7	Low	364.4
10/11/76	60	58	SW	13	0	0	84.7	21.1	440	5.8	37.6	12.7	2.4	Moderate	364.9
10/12/76	66	53	SW	8	0	0	86.2	22.5	444.1	4.7	40	11.2	2	Moderate	366
10/13/76	49	74	NW	12	3	0.15	58.7	16.2	435.4	1	29.6	1.6	0.1	Low	367.8
10/14/76	50	62	SE	12	6	0.22	51.9	10.2	418.7	0.5	19.2	0.5	0	Low	367.6
10/15/76	47	60	NW	7	4	0.12	53.1	7.6	413.9	0.4	14.6	0.3	0	Low	367.9
10/16/76	37	58	NW	9	2	0.11	52.2	5.4	415.1	0.4	10.5	0.3	0	Low	368.2
10/17/76	37	49	NW	10	2	0.05	63.4	5.7	416.3	1.1	11	0.7	0	Low	368.4
10/18/76	35	65	W	3	2	0.11	50	3.8	417.3	0.2	7.4	0.1	0	Low	368.2
10/19/76	40	84	W	3	4	0.17	32.1	1.8	406.1	0	3.6	0	0	Low	368.1
10/20/76	58	83	W	4	0	0	48.2	2.2	409.4	0.2	4.3	0.1	0	Low	367.9

Weather data, CFFDRS codes and indexes
October 21 to October 31, 1976
Michigan DNR Field Office-Seney, MI

Date	Temp	RH	Wind	Wspd	Dur	Rn24	FFMC	DMC	DC	ISI	BUI	FWI	DSR	R	KBDII
	°F	%	Dir	mph	hrs	inch									
10/21/76	35	82	SW	7	5	0.05	50.7	2.3	410.4	0.3	4.5	0.1	0	Low	367.7
10/22/76	34	81	SW	10	9	0.27	31	0.7	387.7	0	1.3	0	0	Low	360.3
10/23/76	35	78		8	0	0	45.8	0.8	388.7	0.2	1.6	0.1	0	Low	360
10/24/76	35	78		7	0	0	56.8	0.9	389.7	0.6	1.8	0.2	0	Low	359.8
10/25/76	36	74	NW	5	3	0.05	57.9	1	390.8	0.5	2	0.2	0	Low	359.7
10/26/76	33	42	N	4	0	0	70.2	1.1	391.6	0.9	2.3	0.3	0	Low	359.6
10/27/76	35	46	N	4	0	0	77.3	1.3	392.6	1.2	2.6	0.4	0	Low	359.3
10/28/76	45	51	S	14	0	0	82.4	1.9	394.6	4.6	3.8	3	0.2	Moderate	359
10/29/76	45	72	S	12	0	0	82.4	2.3	396.6	4	4.4	2.7	0.2	Moderate	359
10/30/76	46	65	NW	5	0	0	82.7	2.8	398.7	2.3	5.4	1.2	0	Low	359.4
10/31/76	44	37	NW	10	0	0	85.9	3.5	400.6	5.3	6.9	4.8	0.4	Moderate	359.7

Palmer Drought Severity Index
April-May 1976

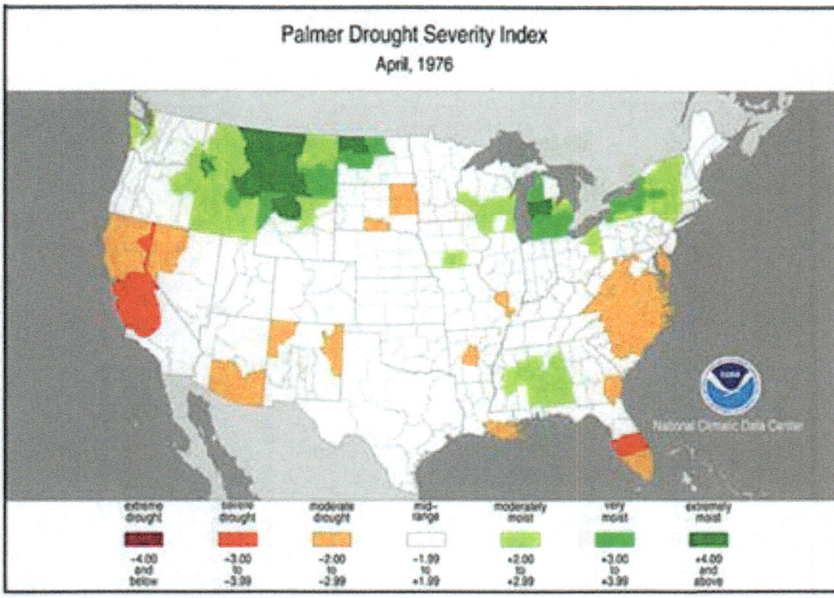

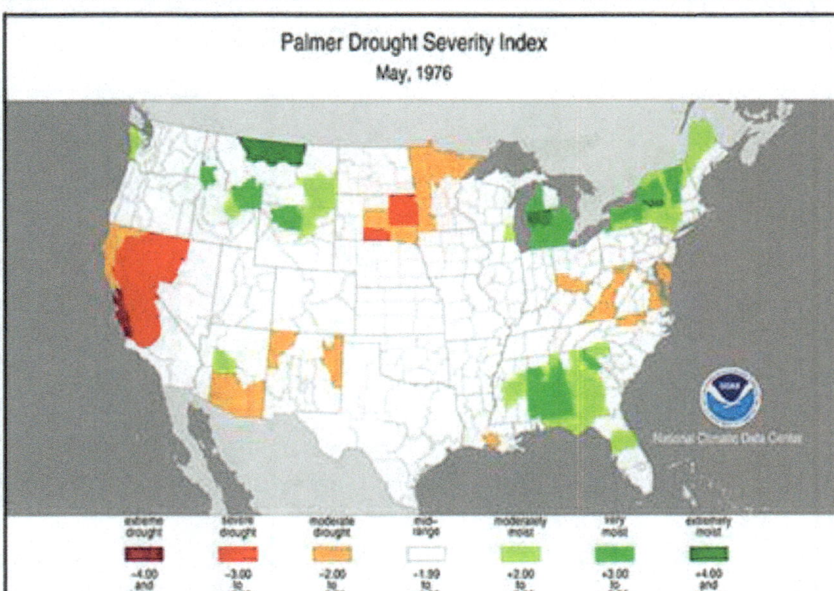

Palmer Drought Severity Index
June-July, 1976

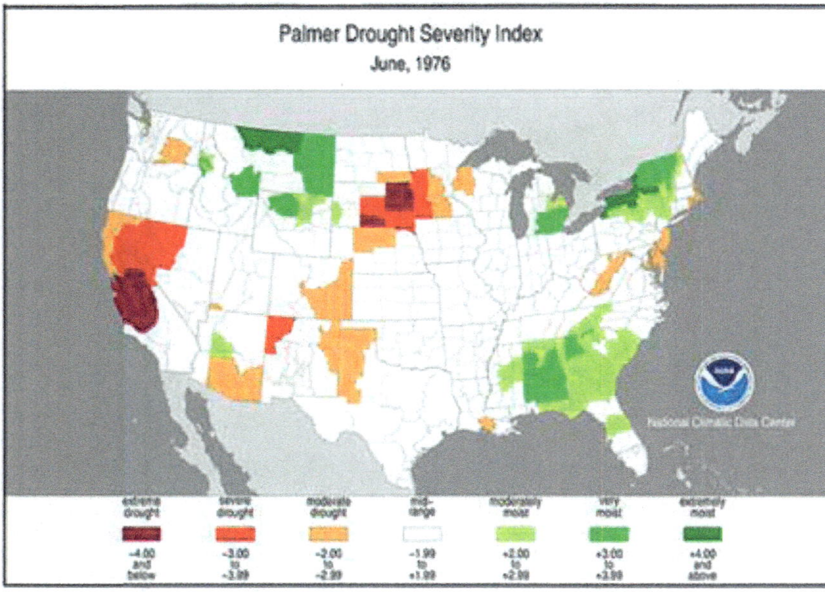

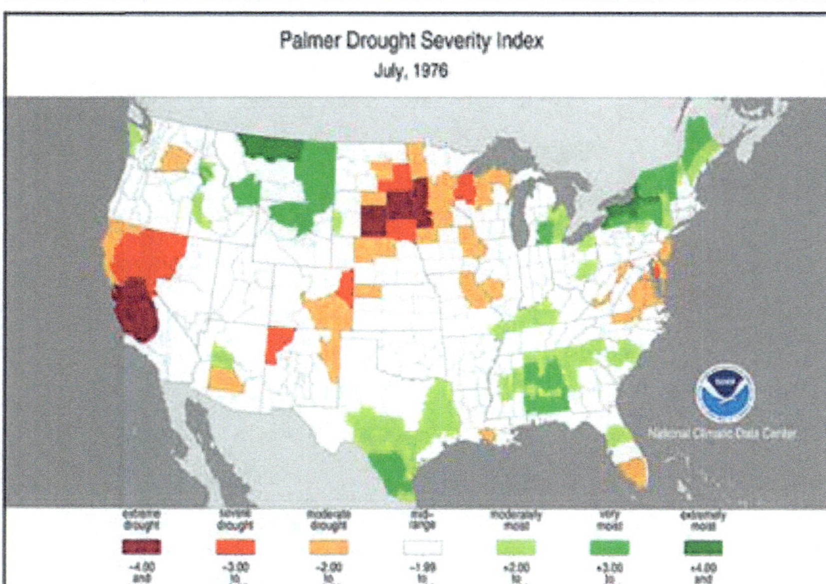

Palmer Drought Severity Index
August-September, 1976

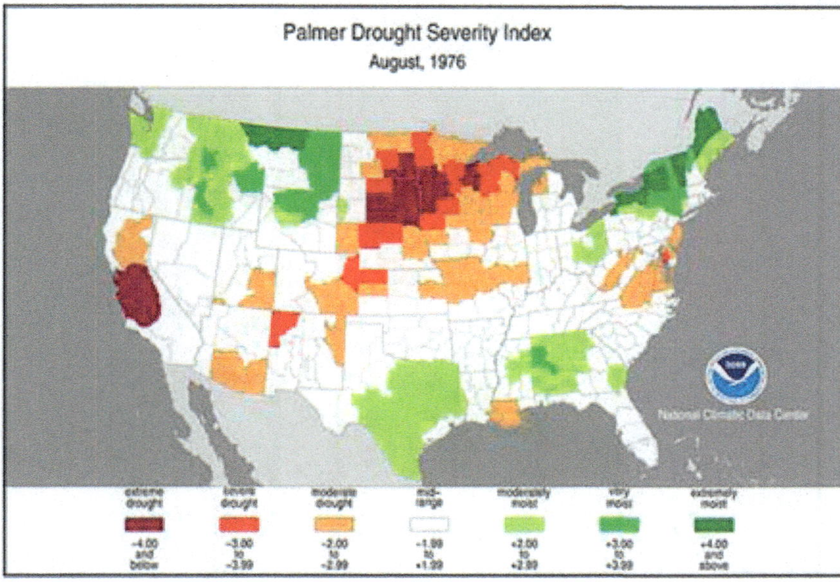

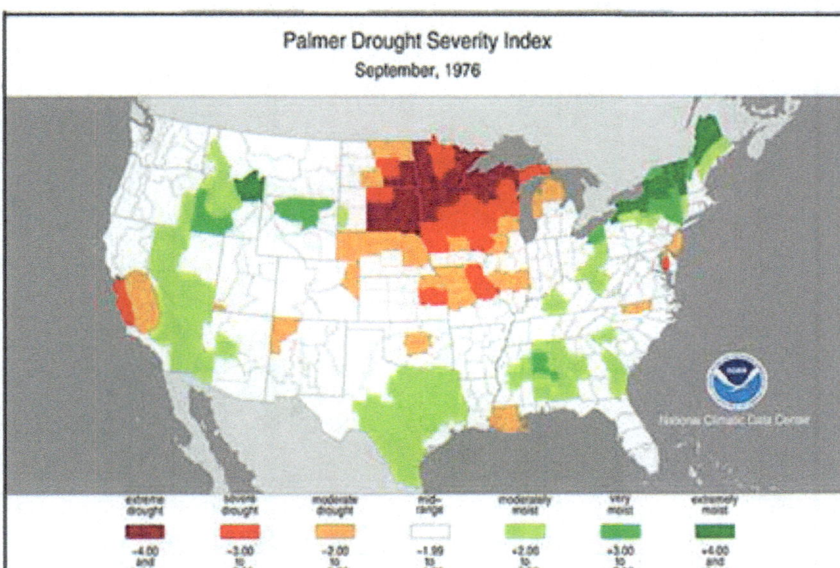

Palmer Drought Severity Index
October-November, 1976

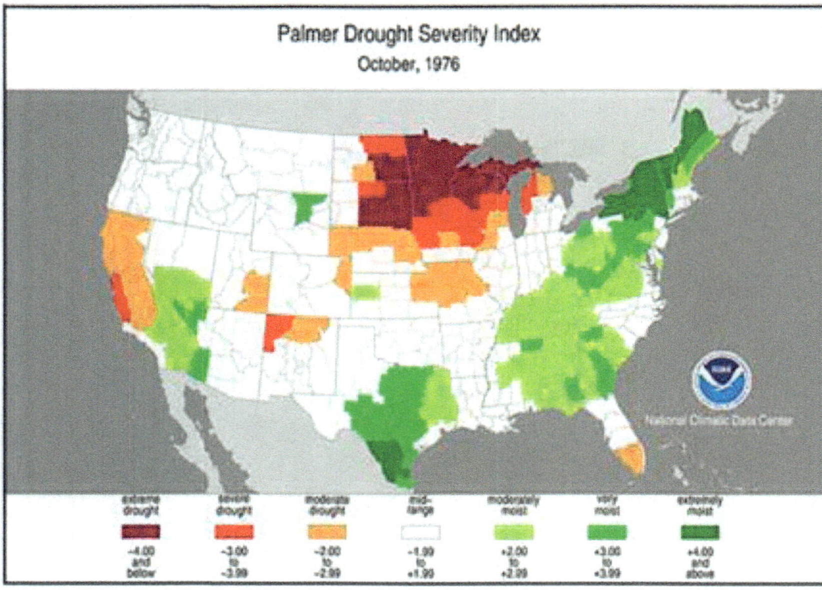

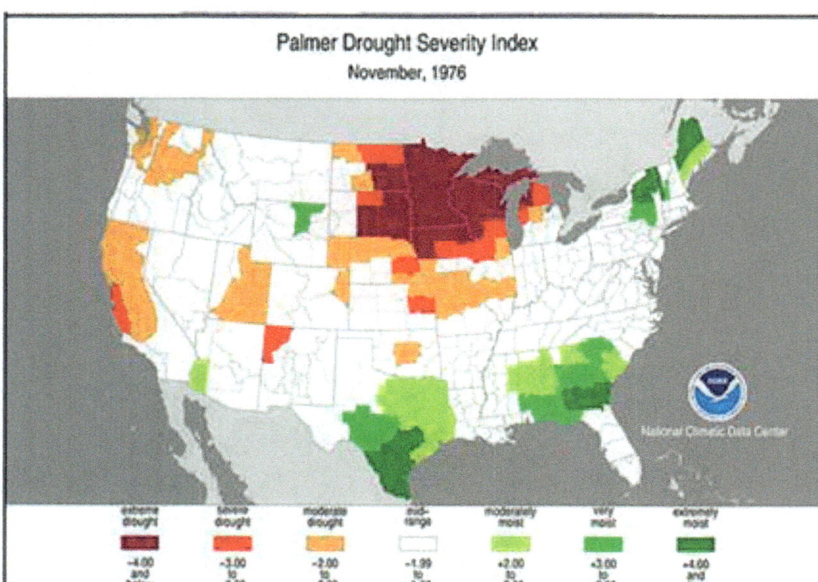

Fire Behavior Tables

2.5.6 C-2, Boreal Spruce

Open, Rate of Spread in ch/hr
Multiply by 1.1 to get feet/min
Divide by 80 to get miles/hour
Divide by 3 to get meters/min
Torching, Active Crown Fire

Intensity Class		Flame Length	FLI kW/m	FLI BTU/ft/sec
1	up to	1	10	3
2	up to	4	500	145
3	up to	8	2000	578
4	up to	12	4000	1156
5	up to	18	10000	2891
6	> than	18	10000	2891

Buildup Index (BUI)

ISI	10	30	50	70	90	110	130	150	170	190
1	0.3	1	1	2	2	2	2	2	2	2
2	0.9	3	4	4	5	5	5	5	5	5
3	2	6	7	8	8	9	9	9	9	9
4	3	8	11	12	12	13	13	13	14	14
5	4	11	15	16	17	18	18	18	19	19
6	5	15	19	21	22	23	23	24	24	24
7	6	18	23	26	27	28	29	29	30	30
8	7	22	28	31	32	34	34	35	35	36
9	8	25	32	36	38	39	40	41	41	42
10	9	29	37	41	43	45	46	47	48	48
11	10	33	42	46	49	51	52	53	54	54
12	11	37	47	52	55	57	58	59	60	61
13	12	41	52	57	61	63	64	66	66	67
14	14	45	57	63	66	69	70	72	73	74
15	15	49	62	68	72	75	77	78	79	80
16	16	52	66	74	78	81	83	84	85	86
17	17	56	71	79	84	87	89	91	92	93
18	18	60	76	84	89	93	95	97	98	99
19	20	64	81	90	95	99	101	103	104	105
20	21	68	86	95	101	104	107	109	111	112
21	22	72	91	100	106	110	113	115	117	118
22	23	75	95	106	112	116	119	121	123	124
23	24	79	100	111	117	122	125	127	129	130
24	25	83	105	116	123	127	130	133	135	136
25	26	86	109	121	128	133	136	139	141	142
26	27	90	114	126	133	138	142	144	146	148
27	28	93	118	131	138	143	147	150	152	154
28	29	97	122	136	144	149	153	155	158	159
29	30	100	127	140	149	154	158	161	163	165
30	31	103	131	145	153	159	163	166	168	170
31	32	106	135	150	158	164	168	171	174	176
32	33	110	139	154	163	169	173	176	179	181
33	34	113	143	158	168	174	178	181	184	186
34	35	116	147	163	172	179	183	186	189	191
35	36	119	151	167	177	183	188	191	194	196
36	37	122	155	171	181	188	193	196	199	201
37	38	125	158	175	185	192	197	201	204	206
38	39	128	162	179	190	197	202	205	208	210
39	40	130	165	183	194	201	206	210	213	215
40	41	133	169	187	198	205	210	214	217	220

Fire Behavior Tables

2.5.4 O-1b, Standing Grass

Open, Rate of Spread in ch/hr

Multiply by 1.1 to get feet/min

Divide by 80 to get miles/hour

Divide by 3 to get meters/min

Intensity Class		Flame Length	FLI kW/m	FLI BTU/ft/sec
1	up to	1	10	3
2	up to	4	500	145
3	up to	8	2000	578
4	up to	12	4000	1156
5	up to	18	10000	2891
6	[illegible]	18	10000	2891

Degree of Curing (%)

ISI	10%	20%	30%	40%	50%	60%	70%	80%	90%	100%
1	0.0	0.0	0.1	0.1	0.2	0.5	1	1	2	2
2	0.0	0.1	0.2	0.4	0.8	2	3	5	6	8
3	0.1	0.2	0.4	0.8	1	3	6	9	12	15
4	0.1	0.3	0.6	1	2	5	9	14	19	23
5	0.1	0.4	0.9	2	3	7	13	20	27	33
6	0.2	0.5	1	2	4	9	18	26	35	44
7	0.2	0.7	1	3	6	11	22	33	45	56
8	0.3	0.8	2	4	7	14	27	41	54	68
9	0.3	1	2	4	8	16	32	48	64	81
10	0.4	1	2	5	9	19	38	56	75	94
11	0.5	1	3	6	11	21	43	64	86	107
12	0.5	1	3	6	12	24	48	73	97	121
13	0.6	2	4	7	14	27	54	81	108	135
14	0.6	2	4	8	15	30	59	89	119	149
15	0.7	2	4	9	16	33	65	98	130	163
16	0.7	2	5	9	18	35	71	106	141	177
17	0.8	2	5	10	19	38	76	114	153	191
18	0.9	2	5	11	21	41	82	123	164	205
19	0.9	3	6	11	22	44	87	131	175	219
20	1	3	6	12	23	46	93	139	186	232
21	1	3	6	13	25	49	98	147	197	246
22	1	3	7	14	26	52	104	155	207	259
23	1	3	7	14	27	54	109	163	218	272
24	1	3	7	15	29	57	114	171	228	285
25	1	4	8	16	30	60	119	179	238	298
26	1	4	8	16	31	62	124	186	249	311
27	1	4	8	17	32	65	129	194	258	323
28	1	4	9	18	34	67	134	201	268	335
29	1	4	9	18	35	69	139	208	278	347
30	2	4	9	19	36	72	143	215	287	359
31	2	4	10	19	37	74	148	222	296	370
32	2	5	10	20	38	76	152	229	305	381
33	2	5	10	21	39	78	157	235	314	392
34	2	5	11	21	40	81	161	242	322	402
35	2	5	11	22	42	83	165	248	330	413
36	2	5	11	22	43	85	169	254	338	423
37	2	5	11	23	44	87	173	260	346	433
38	2	5	12	23	44	88	177	265	354	442
39	2	5	12	24	45	90	181	271	361	452
40	2	6	12	24	46	92	184	276	369	461

MDNR Employee List (Partial)

LAST NAME	FIRST NAME	LOCATION	DIVISION	JOB TITLE
ADAMS	LEROY	NORWAY	FOREST FIRE	AREA FIRE SUPV
AINSLEY	BRIAN	LANSING	FOREST FIRE	STAFF
ALDRICH	ROBERT	D-4		
ALLISON	CLYDE	LANSING	INFO & ED	
ALTO	KEN	MARQUETTE	FOREST FIRE	FIRE OFFICER
AMES	HERBERT		FORESTRY	FORESTER
ANTHONY	JOE	MANISTIQUE	FOREST FIRE	FIRE OFFICER
ARCHAMBEAU	DAVID	ENGADINE	FORESTRY	WORK CREW
AVERY	EARL	HULBERT	FORESTRY	WORK CREW
BABBITT	ROY	MARQUETTE		
BABE	PHIL			
BECKER	CARL	D-5	FOREST FIRE	AREA FIRE SUPV
BELCHER	CHARLES	NEWBERRY	FORESTRY	WORK CREW
BENNETT	DARWIN	LANSING	FOREST FIRE	FIRE MGM SPEC
BENNETT	OWEN	MARQUETTE	ADMIN	REG DIRECTOR
BERGMAN	ALVER			
BERSON	EUGENE	NEWBERRY	TITLE X	
BEZOTTE	JOHN	SAULT STE MARIE		
BOMASTER	CHARLES JR	D-5	FOREST FIRE	FIRE OFFICER
BOMASTER	CHARLES SR	ROSCOMMON	FOREST FIRE	REG SUPV
BORDNER	RODNEY	NEWBERRY	FORESTRY	WORK CREW
BOWERS	CLEO	MIO	FOREST FIRE	FIRE OFFICER
BRICKER	DON	NORWAY	WATERWAYS	EQUIP OPERATOR
BRICKER	EDWAD	NORWAY	WATERWAYS	EQUIP OPERATOR
BROCK	AL	SENEY	FOREST FIRE	FIRE OFFICER
BROZAK	DENNIS	MARQUETTE	ADMIN	BUSINESS EXEC
BRUBAKER	ROBERT	D-8	FOREST FIRE	AREA FIRE SUPV
BRUNET	ROBERT	NEWBERRY	SURVEY	WORK CREW
BRYANT	FREDERICK	NEWBERRY	FORESTRY	WORK CREW
BURNIS	MARCUS	WYMAN NURSERY	FORESTRY	TECHNICIAN
CAMP	LARRY	CRYSTAL FALLS	FOREST FIRE	MECHANIC
CAREY	JAMES	NAUBINWAY	FOREST FIRE	FIRE OFFICER
CARLSON	ELTON	DETOUR	LAW ENF	CONSV OFCR
CARR	MAURICE	DETOUR	FOREST FIRE	FIRE OFFICER
CARSON	KEITH	GAYLORD	FOREST FIRE	MECHANIC
CHALLANCIN	BOB	FELCH	FOREST FIRE	FIRE OFFICER

MDNR Employee List (Partial)

LAST NAME	FIRST NAME	LOCATION	DIVISION	JOB TITLE
CHAMBERLAIN	KURT	NEWBERRY	FORCE ACCT	WORK CREW
CHAMBERS	JOSEPH	NEWBERRY	FORESTRY	WORK CREW
CHAMBERS	JACK	D-4		
CLARK	PATRICK	WAKEFIELD	FOREST FIRE	FIRE OFFICER
CLAUS	ROBERT	MARQUETTE	FOREST FIRE	FIRE OFFICER
COMPEAU	ROBERT	LANSING	FOREST FIRE	DIVISION CHIEF
COTTRELL	KENNETH	D-5		
CRAYCRAFT	JERRY		FOREST FIRE	FIRE OFFICER
DALLAS	JOHN			
DAVIS	GEORGE	NEWBERRY	FORESTRY	WORK CREW
DAVIS	ED	CUSINO		
DELISLE	DANIEL	CUSINO		
DELLIES	WARREN	MARQUETTE	AIR QUALITY	TECHNICIAN
DEVILLEZ	ROBERT	NEWBERRY	FORESTRY	FORESTER
DOVER	RICHARD	NEWBERRY	FORESTRY	WORK CREW
DREW	JOHN	SENEY	FOREST FIRE	FIRE OFFICER
DURLING	BRAD	MARQUETTE	FISHERIES	REG SUPV
ECKELBECKER	GEORGE		LAW ENF	CONSV OFCR
EDWARDS	JOHN	ROSCOMMON	FOREST FIRE	EQUIP OPERATOR
EGGLESTON	SHIRLEY	LANSING	FOREST FIRE	SECRETARY
ENGLISH	GLENN	MCMILLAN	FORESTRY	WORK CREW
ERICKSON	LEO	AUTRAIN	LAW ENF	CONSV OFCR
ERKINTALO	LARRY	NEWBERRY		
FARNSWORTH	DAN	NEWBERRY	FORESTRY	FORESTER
FELDHAUSER	MARK	TRAVERSE CITY		
FOSSITT	JEFF	NEWBERRY	FORCE ACCT	WORK CREW
FRIMODIG	DAVID	MARQUETTE	INFO & ED	PIO
FUNDERBURG	DAN			
GALE	VERN	D-7	FOREST FIRE	FIRE OFFICER
GARROD	STEPHEN	NEWBERRY	FORESTRY	WORK CREW
GILMORE	GARY			
GOUZA	TERRY	NEWBERRY	FORCE ACCT	WORK CREW
GRABANSKI	GODFREY			
GRANGER	DANIEL	NEWBERRY	FORESTRY	WORK CREW
GRAWAY	BERNARD	NEWBERRY	FORESTRY	WORK CREW
GRAY	JAMES	D-7	FOREST FIRE	FIRE OFFICER

MDNR Employee List (Partial)

LAST NAME	FIRST NAME	LOCATION	DIVISION	JOB TITLE
GROH	RAYMOND	NEWBERRY	FORESTRY	WORK CREW
GUSTAFSON	BRUCE	HERMANSVILLE	LAW ENF	CONSV OFCR
GUZINSKI	LEON	EAST LANSING	FOREST FIRE	AREA FIRE SUPV
HAAPALA	CHARLES	NAUBINWAY	LAW ENF	CONSV OFCR
HAATAJA	AARON	CRYSTAL FALLS	FOREST FIRE	FIRE OFFICER
HALBACH	RICHARD	NEWBERRY	TITLE X	
HALL	ERNEST	MARQUETTE		
HAMMERBERG	CELE	NEWBERRY	ADMIN	
HAMMERBERG	MARK	NEWBERRY	TITLE X	
HAMMOND	HAROLD	ST. IGNACE	LAW ENF	CONSV OFCR
HANSON	BARRY	MARQUETTE	FOREST FIRE	FIRE OFFICER
HANSON	ROBERT	NEWBERRY	TITLE X	
HARRINGTON	LAWRENCE	CRYSTAL FALLS	FOREST FIRE	FIRE OFFICER
HARRINGTON	ROBERT	LANSING	ADMIN	PHOTOGRAPHER
HAYES	WILLIAM	D-5	FOREST FIRE	FIRE OFFICER
HEDMARK	GORDON	MARQUETTE	FOREST FIRE	FIRE OFFICER
HEIKKILA	WARREN	ISHPEMING	FORESTRY	TECHNICIAN
HENNIGAR	DAVE	NAUBINWAY	FORESTRY	
HETRICK	ROBERT	NEWBERRY	FORESTRY	WORK CREW
HIELER	ROBERT			
HILL	RANSON	NEWBERRY	LAW ENF	DISTRICT SUPV
HOGUE	KURT	D-4		
HOLLINGSHEAD	ROBERT	MCMILLAN	TITLE X	
HOLSINGER	SAM			
HOSMER	MICHAEL	NEWBERRY	FORCE ACCT	WORK CREW
HOTCHKISS	GEORGE	GOULD CITY	FORESTRY	WORK CREW
HUBBARD	BERNARD	NAUBINWAY	FORESTRY	FORESTER
HUFFMAN	DENNIS	NEWBERRY	FORESTRY	EQUIP OPERATOR
HUMPHREYS	WALLACE	D-6		
HUNTER	DAVID	NEWBERRY	FORESTRY	WORK CREW
ICKES	ROBERT	LANSING		PILOT
JACK	ROBERT			
JACK	LAWRENCE			
JACKOVICH	ALBERT	MARQUETTE	FOREST FIRE	SHOP FOREMAN
JACOBS	KRIS			
JAMSEN	RICHARD	MARQUETTE	FISHERIES	TECHNICIAN

MDNR Employee List (Partial)

LAST NAME	FIRST NAME	LOCATION	DIVISION	JOB TITLE
JARECKI	JOE	NAUBINWAY	FORESTRY	FORESTER
JARUZEL	ROBERT			
JARVIS	FRANK	CHEBOYGAN		
JAYKKA	JOHN	MARQUETTE	FOREST FIRE	MECHANIC
JENSEN	JACK			
JESKE	JAMES	ISHPEMING	WILDLIFE	BIOLOGIST
JOHNSON	ROBERT	NORWAY	FOREST FIRE	FIRE OFFICER
JOHNSON	ELMER	D-4		
JOHNSTON	JOHN	MARQUETTE	FOREST FIRE	EQUIP OPERATOR
JOHNSTON	TERRY	D-7		
JONES	GARY	NEWBERRY	FORESTRY	WORK CREW
JORGENSEN	JACK	ESCANABA	WATERWAYS	SUPV
JOY	GILBERT	D-10	FOREST FIRE	FIRE OFFICER
KAMPPINEN	JOHN	MARQUETTE	ADMIN	REG DISPATCHER
KEINONEN	ELWOOD	NEWBERRY	FORESTRY	WORK CREW
KINGREN	ROBERT	NEWBERRY	FOREST FIRE	AREA FIRE SUPV
KINSEY	BEN	D-11	FOREST FIRE	DISTRICT SUPV
KINTZ	JACK	D-10	FOREST FIRE	AREA FIRE SUPV
KORTE	NORMAN	STEPHENSON	FOREST FIRE	FIRE OFFICER
KOSKI	DENNIS	BARAGA	FORESTRY	TECHNICIAN
KROL	TERRY	GAYLORD		
LAING	PAT	NEWBERRY	TITLE X	MECHANIC
LANGE	JOHN	R-II		
LEMKE	DAVID	NEWBERRY	FORESTRY	WORK CREW
LETTS	MILTON E.			
LEVANEN	JOHN	CHAMPION	FOREST FIRE	FIRE OFFICER
LINTULA	RAY	MARQUETTE	FOREST FIRE	MECHANIC
LITTERAL	RANDY	NEWBERRY	FORESTRY	WORK CREW
LOMAX	WESLEY	ESCANABA	FOREST FIRE	FIRE OFFICER
LONSWAY	RUSSELL	SAULT STE MARIE	FOREST FIRE	FIRE OFFICER
LOUDEN	KEITH	NAUBINWAY	FOREST FIRE	FIRE OFFICER
LUNDQUIST	RALPH	ESCANABA	FORESTRY	TECHNICIAN
LUSK	GREG	MARQUETTE	FOREST FIRE	FIRE MGMT SPEC
MAKI	BILL	MARQUETTE	FOREST FIRE	MECHANIC
MATTHEWS	GLENN	NAUBINWAY		
MCCONNELL	LARRY	TRAVERSE CITY	FOREST FIRE	FIRE OFFICER

MDNR Employee List (Partial)

LAST NAME	FIRST NAME	LOCATION	DIVISION	JOB TITLE
MCGREGOR	JAMES	LANSING	FOREST FIRE	SUPV
MCKIBBIN	RON	LANSING	FOREST FIRE	SUPV
MCMILLAN	JAMES	GWINN	FORESTRY	FORESTER
MCNAMARA	WILLIAM	NEWBERRY	SURVEY	WORK CREW
MCNEAL	THOMAS	NEWBERRY	FORESTRY	WORK CREW
MCNEILL	PATRICK	CURTIS	FORESTRY	WORK CREW
MEAD	KENNETH	NEWBERRY	FORESTRY	WORK CREW
MICHAEL	FOSTER	NEWBERRY	FORESTRY	WORK CREW
MILLER	THOMAS	MCMILLAN	FORESTRY	WORK CREW
MILLER	WILLIAM	CUSINO		
MILLS	BLANCHARD	LANSING	INFO & ED	PHOTOGRAPHER
MITCHELL	ROBERT	MCMILLAN	TITLE X	
MOERDYK	ERROLL	NEWBERRY		RADIO TECH
MONCK	ROBERT	D-4		
MULLENDORE	WILLIAM	LANSING	INFO & ED	PIO
MURPHY	TIMOTHY			
MYERS	DALE		LAW ENF	CONSV OFCR
NATHAN	LAWRENCE	D-5	FOREST FIRE	FIRE OFFICER
NOASCONI	JIM	STEPHENSON	FOREST FIRE	AREA FIRE SUPV
NUNEMAKER	RICK	NEWBERRY	FORCE ACCT	WORK CREW
OGLE	DON	D-4		
OLIVERIUS	WILLIAM	CRYSTAL FALLS	FOREST FIRE	DISTRICT SUPV
OLSEN	VERNARD	NEWBERRY	FOREST FIRE	FIRE OFFICER
OSBORNE	HAROLD			
OSWALD	ANTHONY	D-5	FOREST FIRE	FIRE OFFICER
OWEN	KEITH	NEWBERRY	TITLE X	
PALO	CHARLES	STEPHENSON	FOREST FIRE	FIRE OFFICER
PALUDA	MICHAEL	NEWBERRY	FOREST FIRE	DISTRICT SUPV
PARSONS	JAMES	D-6	FOREST FIRE	FIRE OFFICER
PATTEN	JACK	NEWBERRY	TITLE X	
PERRY	MATTHEW	NEWBERRY	FORCE ACCT	WORK CREW
PERRY	HENRY	NEWBERRY	TITLE X	
PERRY	LEO	CUSINO		
PETERSON	RICHARD	MORAN	LAW ENF	CONSV OFCR
PITONQUIT	STEVEN	NEWBERRY	FORESTRY	WORK CREW
PRICE	KERRY	GWINN	FORESTRY	

MDNR Employee List (Partial)

LAST NAME	FIRST NAME	LOCATION	DIVISION	JOB TITLE
PUDELKO	JAMES	DETOUR	FOREST FIRE	FIRE OFFICER
REID	PAUL	ROSE LAKE		
REYNOLDS	BILL	MARQUETTE	FOREST FIRE	MECHANIC
RICE	HOWARD	BARAGA	FOREST FIRE	DISTRICT SUPV
ROBERTSON	JOHN	NEWBERRY	FISHERIES	DISTRICT SUPV
ROWELL	LYLE	GRAND MARAIS	LAW ENF	CONSV OFCR
ROZYCKI	RONALD	KALKASKA	FOREST FIRE	FIRE OFFICER
RUBICK	LAWRENCE	NEWBERRY	FORESTRY	TECHNICIAN
RUBICK	LAWRENCE JR	NEWBERRY	FORCE ACCT	WORK CREW
RYAN	STAN	D-8		
SADLER	DARRELL			
SAFFORD	TOM	ESCANABA	FOREST FIRE	FIRE OFFICER
SCHLUETER	PAUL	D-4	LAW ENF	CONSV OFCR
SCHMALZ	TED	WAKEFIELD	FOREST FIRE	FIRE OFFICER
SCHMIDT	CAROL	NEWBERRY	ADMIN	RADIO OP
SCHNURER	GARY	SENEY	FOREST FIRE	FIRE OFFICER
SCHULTZ	ARLYN			
SCHUSTER	JOE	ESCANABA	FOREST FIRE	FIRE OFFICER
SEVARNS	JAMES	NEWBERRY	FORESTRY	WORK CREW
SHAPTON	WARREN	LANSING	ADMIN	DEPUTY DIRECTOR
SHIRTZ	CHARLES	MARQUETTE		ELECTRICIAN
SILER	DELL	CRYSTAL FALLS	FISHERIES	
SIMMONS	MAX	NEWBERRY	TITLE X	
SKINNER	BILL	D-4		
SKOVERA	MATTHEW	BOYNE CITY	FOREST FIRE	FIRE OFFICER
SMITH	A.J.	WYMAN NURSERY	FORESTRY	TECHNICIAN
SNYDER	JOHN	MARQUETTE	FOREST FIRE	PILOT
SNYDER	JOHN D	MARQUETTE		
SPALDING	DAVE	CUSINO	FORESTRY	
SQUIBB	HARRY	CRYSTAL FALLS	WILDLIFE	BIOLOGIST
ST. ONGE	WILLARD	TROUT LAKE	LAW ENF	CONSV OFCR
ST. OURS	DUANE	D-4	FOREST FIRE	FIRE OFFICER
ST. OURS	HARLEY	NAUBINWAY	FORESTRY	TECHNICIAN
STAHL	CARL	D-5	FOREST FIRE	FIRE OFFICER
STARK	HAROLD	D-4	LAW ENF	CONSV OFCR
STARK	TERRY	NEWBERRY		

MDNR Employee List (Partial)

LAST NAME	FIRST NAME	LOCATION	DIVISION	JOB TITLE
STEPHENSON	RONALD			
STEVENS	DON	MARQUETTE	AIR QUALITY	SUPV
STOWELL	ROBERT	LANSING	FOREST FIRE	SUPV
SUCH	STEVE	ROSCOMMON	FOREST FIRE	EXP STATION SUPV
SZABO	GEORGE	ESCANABA	FOREST FIRE	MECHANIC
SZYMANSKI	WALTER	D-5	FOREST FIRE	FIRE OFFICER
TANNER	HOWARD	LANSING	ADMIN	DIRECTOR
TAYLOR	CECIL	NEWBERRY	LAW ENF	CONSV OFCR
TENNIS	TIM	D-4		
TESSIER	GEORGE	NEWBERRY	FORESTRY	WORK CREW
TRUDEAU	DAVID	NEWBERRY	FORESTRY	WORK CREW
TUCKER	CLIFFORD	NEWBERRY	FORESTRY	WORK CREW
TUOVILA	DAVID	CRYSTAL FALLS	FOREST FIRE	FIRE OFFICER
VALENTINE	RICHARD	D-5	FOREST FIRE	FIRE OFFICER
VANSETTERS	BENNIE	NEWBERRY	TITLE X	
VANSIMAEYS	FRED	ROSCOMMON	FOREST FIRE	PILOT
VENEBERG	BRUCE	CUSINO	FORESTRY	FORESTER
VERCRUYSSE	VICTOR	PORTER RANCH	FOREST FIRE	FIRE OFFICER
VERRETT	CARL	ENGADINE	FORESTRY	WORK CREW
VERTICAN	CARL			
VETORT	EDWARD	ESCANABA	FOREST FIRE	DISTRICT SUPV
WALKER	JOHN	MANISTIQUE	LAW ENF	CONSV OFCR
WALLACE	WILFRED	WYMAN NURSERY	FORESTRY	TECHNICIAN
WALSTROM	LESLIE	CURTIS	LAW ENF	CONSV OFCR
WALTERS	CLINTON	KALKASKA	FOREST FIRE	AREA FIRE SUPV
WEBER	DAVID	NEWBERRY	FORESTRY	WORK CREW
WEBSTER	AL	NEWBERRY	FORESTRY	AREA FORESTER
WESCOTT	BOB	FELCH	FOREST FIRE	FIRE OFFICER
WHEELER	MORRIS	NORWAY	FOREST FIRE	FIRE OFFICER
WHITMER	GARY	NEWBERRY	FORCE ACCT	WORK CREW
WIGGINS	GERALD	NEWBERRY		
WILCOX	DANIEL	CRYSTAL FALLS	FORESTRY	FORESTER
WILSON	RONALD	LANSING	FOREST FIRE	SUPV
WOOD	ROBERT	NEWBERRY	TITLE X	
WOODWARD	RALPH	GAYLORD	FOREST FIRE	MECHANIC
YAEGER	ROBERT			

MDNR Employee List (Partial)

LAST NAME	FIRST NAME	LOCATION	DIVISION	JOB TITLE
YORK	OVID	GWINN	WILDLIFE	EQUIP OPERATOR
YOUNG	GEORGE	MANISTIQUE	FOREST FIRE	AREA FIRE SUPV
ZIMMERMAN	JOHN	D-4	LAW ENF	CONSV OFCR

Perimeter Map Showing Origin

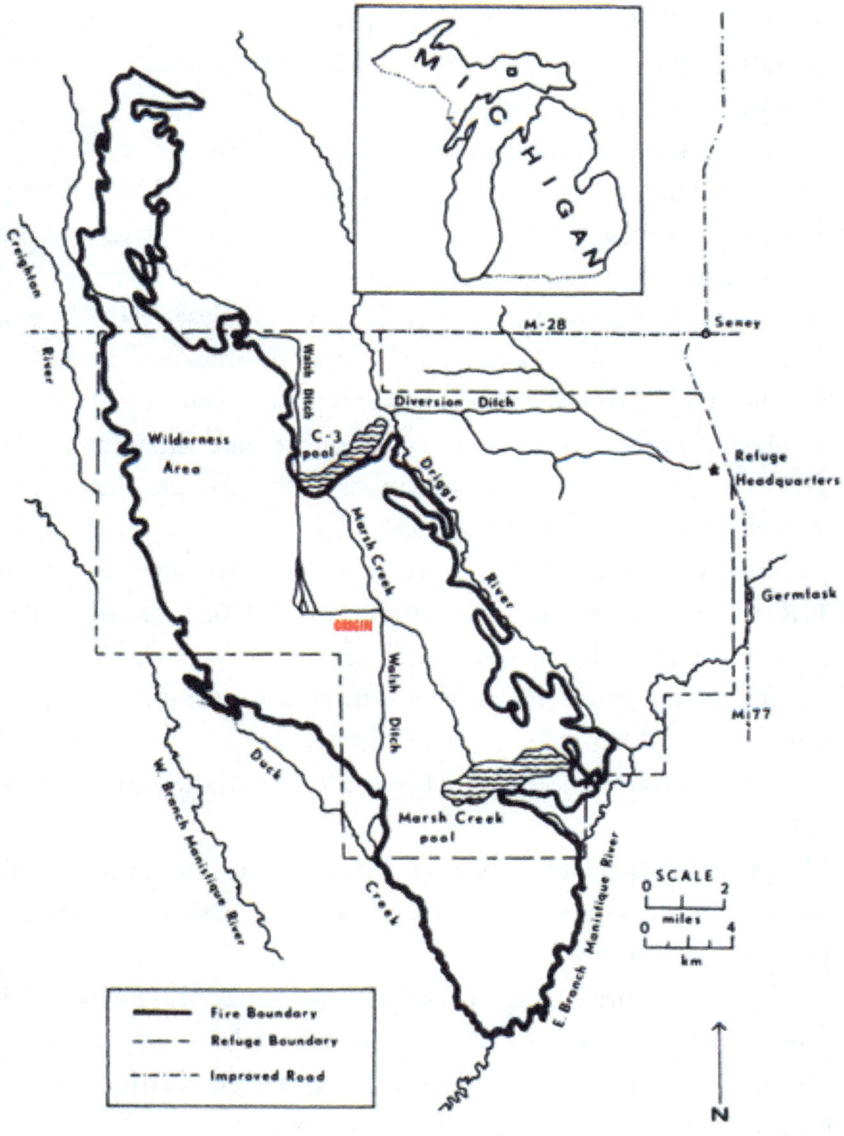

Original map created by Stanley H. Anderson in Resource Publication 146, *Effects of the 1976 Seney National Wildlife Refuge Wildfire on Wildlife and Wildlife Habitats*, F&WS, USDI, July 1982, page 2, edited February 2023, showing Walsh Ditch and the fire origin.

ACKNOWLEDGEMENTS

It took much too long to complete this project but it couldn't have been done without the assistance of many people. My thanks goes especially to the following and also to anyone I've forgotten who was instrumental in this project.

Bill Rockwell, for his keen eye and expertise in editing.

Mike Paluda, fire boss on the fire in 1976 and the MDNR State fire supervisor who is now enjoying his retirement, for helping me remember details that had escaped my memory.

Cory Luoto, Mobile Technology Training Analyst, MDNR, who serves on national fire teams as a Geographic Information System Specialist, for creating an extremely useful progression map of the fire.

Robert Ziel for helping me to better understand fire weather, fire behavior and the Canadian Fire Danger Rating System through his extensive knowledge and experience.

Celeste Chingwa, UP Resource Protection Manager, Michigan DNR, for providing weather data from 1976 and for answering many questions about present fire management operations.

Tom Proulx, retired Forest Fire Officer, for several news articles from the period that helped provide more depth.

Jeff Noble, retired Forest Fire Officer for moral support and technical expertise.

Dr. Randall Schaetzl, Michigan State University, for graciously providing permission to use images from his Physical Geography-classroom presentations.

Joshua Haen, Fire Management Specialist, Seney National Wildlife Refuge, for providing a copy of the refuge fire plan.

John Clearwater, Public Affairs Officer, Angeles National Forest for his review of the story written by the hotshots.

Especially to my better half, Sandra, who endured my time away from her when I was writing this book.

REFERENCES

A Little History of Seney, Michigan in the Upper Peninsula, http://www.exploringthe-north.com/seneymi/history.html

Amadeo, Kimberly, *How Wildfires Impact the Economy*, https://www.thebalance.com/wildfires-economic-impact-4160764, accessed June 11, 2021

Anderson, Stanley H., Resource Publication 146, *Effects of the 1976 Seney National Wildlife Refuge Wildfire on Wildlife and Wildlife Habitats*, Fish and Wildlife Service U.S. Department of the Interior, July 1982

Baron, Ethan, *New Technology Aims to Help Fight California Wildfires,* https://www.govtech.com/news/new-technology-aims-to-help-fight-california-wildfires, Mercury News, June 28, 2021, accessed May 3, 2022

Bohnak, Karl, *So Cold a Sky: Upper Michigan Weather Stories.* United States: Cold Sky Publishing, 2006

Brain, C., *The three ages of fire,* Nature, 414, 584, https://doi.org/10.1038/414584a, 2001

Brock, Hannah, Great Lakes Echo, *Great Lakes wildfires could double,* https://greatlakesecho.org/2021/10/19/great-lakes-wildfires-could-double/, October 19, 2021

Charter Township of Chocolay-History, *Then and Now*, https://www.chocolay.org/documents/townshiphistory.pdf, June 2008, p. 18

Climate Narrative for Schoolcraft County, Michigan, Prepared by the Natural Resources Conservation Service National Water and Climate Center, Portland, Oregon, www.wcc.nrcs.usda.gov/legacy/ftp/support/climate/soil-nar/mi/schoolcraft.doc

Corace, Greg; Goebel, Charles; Kashian, Dan, Seney National Wildlife Refuge, *Reconciling Yellow Rail Habitat Use and Landscape Dynamics at Seney National Wildlife Refuge*, https://slidetodoc.com/seney-national-wildlife-refuge-reconciling-yellow-rail-habitat/

Crumley, Bruce, *Firefly (really) heavy-lift drone boasts mega-endurance, too*, https://dronedj.com/2022/01/04/firefly-really-heavy-lift-drone-boasts-mega-endurance-too/, January 4, 2022, accessed May 1, 2022

Daily Press, Escanaba, MI, *"Arson Suspected"*, September 23, 1976

Daily Press, Escanaba, MI, *Threatening Fire Contained*, September 20, 1976

Department of Natural Resources, *Partial List of State Employees*, District 4 records, Newberry, MI October 11, 1976

Drone Amplified Editor, *U.S. Forest Service and Drone Amplified partner to drive the search for fire-fighting drones*, https://droneamplified.com/u-s-forest-service-and-drone-amplified-partner-to-drive-search-for-domestic-fire-fighting-drones/?v=7516fd43adaa, September 9, 2020, accessed May 1, 2022

Duck Lake Fire, Wikipedia, https://en.wikipedia.org/wiki/Duck_Lake_Fire, 16 August 2020

Eastern Area Coordination Center, https://gacc.nifc.gov/eacc/admin/about_us/about us.htm, accessed June 11, 2021

Eastern Upper Peninsula Regional State Forest Management Plan Section 2, Forest History, https://www2.dnr.state.mi.us/publications/pdfs/forestslandwater/RSFMP/EUP/EUPSection2.pdf

Escanaba Daily Press, *"Feds Pay State Nearly $1 Million for Seney Fire"*, June 7, 1977

Escanaba Daily Press, *"Fire hearing slated"*, September 4, 1976

Escanaba Daily Press, *"Seney hearing set"*, June 24, 1977

Guinness World Records, *First Wildfire*, https://www.guinnessworldrecords.com/world-records/first-wildfire, accessed May 4, 2022

Foster, Z.C., *Soil Survey, Schoolcraft County, Michigan*, United States Bureau of Chemistry and Soils, United States Bureau of Plant Industry, Soils and Agricultural Engineering, United States Soil Conservation Service, United States Natural Resources Conservation Service, p. 38, (retrieved February 1, 2018) from Google Free E-books: https://books.google.com/books?id=QfjwAAAAMAAJ&pg=PA2&source=gbs_selected_pages&cad=3#v=onepage&q&f=false

Geographic Area Coordination Centers, https://gacc.nifc.gov/admin/disclaimer/disclaimer.htm, accessed June 11, 2021

Goebel, P. Charles, et. al., *Fire History Study Reveals Surprises about Mixed-Pine Ecology in Eastern Upper Michigan*, Fire Science Brief Issue 132, JFSP Project Number 05-2-1-86, https://www.firescience.gov/projects/briefs/05-2-1-86_FSBrief132.pdf, April 2011

Gogan, Peter J. P., and Cochrane, Jean Fitts, *Restoration of Woodland Caribou to the Lake Superior Region*, U.S. National Park Service Publications and Papers. 11. http://digitalcommons.unl.edu/natlpark/11, 1994

Goodrich, Marcia, *Back to Nature*, Michigan Tech Magazine, Volume 50, Number 1, Michigan Technological University, Spring 2013,

Haines, Donald A., Johnson, Von J., and Main, William A., *An Assessment of Three Measures of Long-Term Moisture Deficiency Before Critical Fire Periods*, https://www.nrs.fs.fed.us/pubs/rp/rp_nc131.pdf, 1976

Haines, Donald A., Sando, Rodney W., *Climatic Conditions Preceding Historically Great Fires in the North Central Region,* North Central Forest Experiment Station Forest Service U.S. Department of Agriculture, Research Paper NC-34, 1969

Harper, Francis. *The Barren Ground Caribou of Keewatin*. Lawrence, Kan.: University of Kansas, 1955

Heckman, Kerry, PIO, Michigan.gov, *Firefighters have Blue Lakes Fire 75% contained, drones deployed to map hot spots,* https://www.michigan.gov/dnr/about/newsroom/releases/2022/05/16/blue-lakes-fire-75-percent-contained, May 16, 2022

Hupy, Christina M., and Yansa, Catherine H., *The Last 17,000 Years of Vegetation History,* https://msu.edu/~yansa/2009a-Hupy%20&%20Yansa-CH%207-MI%20Book%20 Chapter.pdf, 2009

Jobst, Jack, *Hemingway in Seney,* Michigan Historical Magazine, November/December 1990, https://www.cmich.edu/library/clarke/ResearchResources/Michigan_Material_Statewide/Hemingway_in_Michigan/Pages/Hemingway-in-Seney.aspx

Johnson, Coralie Cederna, *A Shadowy Saga of Seney*, reprinted from Peninsula People Magazine, https://wildwoodpress.org/a-shadowy-saga-of-seney/, 2017

Kennedy, Caitlyn, *Risk of very large fires could increase sixfold by mid-century in the US,* Climate. Gov, https://www.climate.gov/news-features/featured-images/risk-very-large-fires-could-increase-sixfold-mid-century-us, published August 26, 2015, updated July 2, 2021, accessed April 11, 2022

Leefers, Larry A., Dickman, Donald I., *The Forests of Michigan*, Revised Ed., United States: University of Michigan Press, 2016

List of Michigan Wildfires, Wikipedia, https://en.wikipedia.org/wiki/List_of_Michigan_wildfires, accessed January 22, 2021

Little, Jane Braxton, *Firefighting Robots Go Autonomous*, Scientific American, https://www.scientificamerican.com/article/firefighting-robots-go-autonomous/ October 29, 2021, accessed May 4, 2022

Loomis, Bill, *Shanty Boys, River Hogs and the Forests of Michigan*, Business Section, The Detroit News, Apr 11, 2012. http://blogs.detroitnews.com/history/2012/04/11/shanty-boys-river-hogs-and-the-forests-of-michigan/

Losey, Elizabeth Brown, *Seney National Wildlife Refuge, Its Story*, Lake Superior Press, Marquette, MI, 2003

Luoto, Cory, Michigan Department of Natural Resources, *Personal Communication*, February 19, 2018

Marsh, Dawn S., *The Evolution of Land and Fire Management at Seney National Wildlife Refuge, From Game to Ecosystem Management*, https://www.fws.gov/uploadedFiles/Marsh_2013_fire_conceptual_changes.pdf, December 2013

McCay, Jim, Government Technology, *Mesh Network Creating Situational Awareness During Wildfires*, https://www.govtech.com/em/disaster/mesh-network-creating-situational-awareness-during-wildfires.html, April 5, 2019

McLennan, Amber, RightMesh, *What is Mobile Mesh Networking?*, https://medium.com/rightmesh/what-is-mobile-mesh-networking-964732814943, March 1, 2019, accessed May 2, 2022

Melton, Mike, *The Keetch-Byram Drought Index: A Guide to Fire Conditions and Suppression Problems*, Fire Management Notes; Vol 50, No 4, 1989, p. 30-33

Michelle, *Pillar Point, Alert Fire, and Smoke Point Use Harbor Camera Now Pointing at El Granada Hills*, https://www.coastsidebuzz.com/pillar-point-alert-fire-and-smoke-point-use-harbor-camera-now-pointing-at-el-granada-hills/, August 3, 2021, accessed April 30, 2022

Michigan Wildfires, Wikipedia, https://en.wikipedia.org/wiki/List_of_Michigan_wildfires, accessed January 22, 2021

Mitchell, J.A., and Robson, D., *Forest Fires and Forest Fire Control in Michigan*, Michigan Department of Conservation, In Cooperation with United States Department of Agriculture Forest Service, Lake States Forest Experiment Station, with the University of Minnesota, University Farm, St. Paul, MN, 1950

MPSCS-Michigan's Public Safety Communications System, *Inside MPSCS*, https://www.michigan.gov/mpscs/inside, accessed June 10, 2021

Myers, John; Northland Outdoors, *Last Lake Superior Caribou Surviving on Their New Island Homes*, http://www.northlandoutdoors.com/2020/04/06/last-lake-superior-caribou-surviving-on-their-new-island-homes/, April 6, 2020

National Park Service, U.S. Department of the Interior, Pictured Rocks National Lakeshore, *Logging History*, https://www.nps.gov/piro/learn/historyculture/logging-history.htm, Last updated: November 17, 2021, accessed May 4, 2022

National Wildfire Coordinating Group, *Fire Behavior Prediction (FBP) System*, https://www.nwcg.gov/publications/pms437/cffdrs/fire-behavior-prediction-system, accessed May 18, 2022

Neumann, Brad, *Climate Change and Wildfire in the Great Lakes Region*, Michigan State University, Extension Bulletin E-3277, August 2015

News-Herald, Michigan Department of Community Health, Michigan Department of Natural Resources, and Michigan Department of Transportation, *A view from above,* https://www.thenewsherald.com/2020/06/28/a-view-from-above-michigan-dnr-explores-drone-technology/, Published: June 28, 2020, updated: June 17, 2021, accessed May 1, 2022

NOAA-National Centers for Environmental Information, *Historical Palmer Drought Indices*, https://www.ncdc.noaa.gov/temp-and-precip/drought/historical-palmers/, accessed May 4, 2022

Paleontology in Michigan, Wikipedia, https://en.wikipedia.org/wiki/Paleontology_in_Michigan November 13, 2021, accessed May 4, 2022

Pesgraves, David, The Last Mile, *3 mesh network use cases for wildland fire deployment,* https://thelastmile.gotennapro.com/3-mesh-network-use-cases-for-wildland-fire-deployment/, April 7, 2022, accessed May 2, 2022

Pierce, Charles S., *Eleventh Bicentennial Report of the State Game, Fish, and Forestry Warden*, Winkoop, Hollenbeck Crawford Company, State Printer, Lansing, Michigan, 1909

Fish & Wildlife Service, Field Notes Entry, *Pine Creek North Wildfire-Seney National Wildlife Refuge,* https: //www.fws.gov/fieldnotes/regmap.cfm?arskey=32575, June 2012

Prell, Adam, RedZone, *How Cameras Benefit Fire Fighting Efforts,* https://www.redzone.co/2022/04/29/how-cameras-benefit-fire-fighting/?utm_campaign=External%20Blog&utm_medium=email&_hsmi=211623536&_hsenc=p2ANqtz-8nQwkoerHMDI-SgK-d1pGSioW43xzy2OxHknsekVGf93PFj4JfLfIaF1ienpLiMB4qwgQ-V9-677UtoUtbABvY-sWM_B_Q&utm_content=211623536&utm_source=hs_email, April 29, 2022, accessed April 30, 2022

President's Council of Advisors on Science and Technology, *Report to the President, Modernizing Wildland Firefighting to Protect Our Firefighters, Executive Office of the President*, https://www.whitehouse.gov/wp-content/uploads/2023/02/PCAST_Wildfires-Report_Feb2023.pdf, February, 2023, Accessed March 1, 2023

Pulliam, Jim, Chairman, *Report of Walsh Ditch Fire Board of Review,* October 1, 1976

Pyne, Stephen J., *Between Two Fires*, A Fire History of Contemporary America, The University of Arizona Press, Tucson, The Arizona Board of Regents, 2015

Rossi, Marcello, *From Supercomputers to Fire-Starting Drones, These Tools Help Fight Wildfires,* Smithsonian Magazine, https://www.smithsonianmag.com/innovation/from-supercomputers-to-fire-starting-drones-these-tools-help-fight-wildfires-180978421/, August 20, 2021, accessed May 3, 2022

Schaetzl, Dr. Randall, *Geography of Michigan & the Great Lakes*, GEO 333, Part II. Physical Geography, Spring Semester, 2005, Michigan State University, http://geo.msu.edu/extra/geomich/part two-B.html, and http//geo.msu.edu/extra/geomich/glacial-lake_algonq.html

Schradin, Ryan, *Four considerations for scaling mesh networks on the fireline*, https://thelastmile.gotennapro.com/four-considerations-for-scaling-mesh-networks-for-wildfire-response/, October 19, 2020, accessed May 2, 2022

Seney National Wildlife Refuge-General Brochure, https://www.fws.gov/uploadedFiles/Region_3/NWRS/Zone_2/Seney_Complex/Seney/Sections/Newsletters/SNYGeneral-Brochure.pdf

Serba, John, mLive.com, *Meet Michigan's Most Bizarre Prehistoric Creatures,* https://www.mlive.com/entertainment/2017/08/michigans_wildest_prehistoric.html, Posted Aug 13, 2017, updated May 20, 2019

Sleeper Lakes Fire, Wikipedia, https://en.wikipedia.org/wiki/Sleeper_Lakes_Fire, last update March 17 2022, accessed May 9, 2022

Sniegowski, Walt; Herrand, David; McGeragle, Mike; Vrandenberg, Mike; Satern, Monte; Wrench, Rod, *Angeles National Forest Alumni Newsletter 40 Year Anniversary Edited and Arranged,* Volume 1, No. 3, Summer 2010

Sodders, Betty, *Michigan on Fire*, Thunder Bay Press, Holt, MI, 1997

Spiroff, Prof. Kiril, Geologic History of Michigan, July 1-2-3-4, 1960, Mich. College of Mining & Technology Midwest Federation of Mineralogical & Geological Societies Twentieth Annual Field Trip Convention, *"Souvenir Bulletin of the Jaspilite"*

Stavros, Natasha, The Conversation, *Bringing tech innovation to wildfires: 4 recommendations for smarter firefighting as mega-fires menace the US*, https://theconversation.com/bringing-tech-innovation-to-wildfires-4-recommendations-for-smarter-firefighting-as-megafires-menace-the-us-162178, June 8, 2021, accessed May 4, 2022

Stearns, Forest W., *History of the Lake States Forests: Natural and Human Impacts,* In: Vasievich, J. Michael; Webster, Henry H., eds., Lake States Regional Forest Resources Assessment: Technical Papers. Gen. Tech. Rep. NC-189. St. Paul, MN: U.S. Department of Agriculture, Forest Service, North Central Forest Experiment Station: 8-29, 1997

Stewart, Luke, The Last Mile, *Situational awareness? Thanks to ATAK, there's an app for that!*, https://thelastmile.gotennapro.com/situational-awareness-thanks-to-atak-theres-an-app-for-that/, February 4, 2021, accessed May 2, 2022

The Evening News, Sault Ste. Marie, MI, October 12, 1976

The Mining Journal, Marquette, MI, *Seney Stretch and Seney wildlife refuge*, July 15, 2014

The Mining Journal, Marquette, MI, *"DNR to bill Uncle Sam"*, August 25, 1976

The Mining Journal, Marquette, MI, *"Investigation Required"*, p. A-8, August 25, 1976

The Mining Journal, Marquette, MI, *"Seney fire near end"*, front page, August 31, 1976

The Mining Journal, Marquette, MI, *"Congressional hearing"*, September 4, 1976

The Mining Journal, Marquette, MI, *"Let burn hassle rages on"*, September 16, 1976

The Mining Journal, Marquette, MI, *"Seney fire handling to be reviewed; Kleppe will study 'let-burn' policy"*, September 23, 1976

The Mining Journal, Marquette, MI, *"Sec Kleppe confirms feds will pay for Seney fire"*, p. 3-A, September 23, 1976

The Mining Journal, Marquette, MI, *"A broken heart"*, November 12, 1976

The Mining Journal, Marquette, MI, December 13, 1976, p. A-2

The People History, *What Happened in 1976, Important News and Events*, Key Technology and Popular Culture, http://www.thepeoplehistory.com/1976.html, accessed May 4, 2022

The U.P.'s Natural Landmark – *Strangmoor Bog*-The Unique Formation of Seney's Natural National Landmark Conservation Gateway, Nature Conservancy, https://www.conservationgateway.org/ConservationByGeography/NorthAmerica/UnitedStates/michigan/projects/Documents/Landscape%20Stewardship%20Stories/14.StrangmoorBog.pdf posted 7/18/2017, accessed May 4, 2022

The White House, Fact Sheet: *The Biden-Harris Administration Acts to Address the Growing Wildfire Threat*: https://www.whitehouse.gov/briefing-room/statements-releases/2021/06/30/fact-sheet-the-biden-harris-administration-acts-to-address-the-growing-wildfire-threat/, June 30, 2021

Urbanek, Richard P. and Bookhout, Theodore A., *"Nesting of Greater Sandhill Cranes on Seney National Wildlife Refuge"*, North American Crane Workshop Proceedings. 325. http://digitalcommons.unl.edu/nacwgproc/325, 1992

U.S. Fish & Wildlife Service, *Fire Management*, http://www.fws.gov/fire/who_we_are/history.shtml, last updated May 28, 2014

U.S. Fish & Wildlife Service, *National Wildlife Refuge System*, https://www.fws.gov/refuges/airquality/aris/sene/index.html

U.S. Fish and Wildlife Service, *Role of Fire in Alaska*, https://www.arlis.org/docs/vol1/83599936/firetriangle.html, USF&WS Alaska Region Division of Information Resource Management (IRM) and External Affairs (EA)

University of Utah, Mesowest, *Great Lakes Fire &Fuels*, https://glff.mesowest.org/, accessed, May 4, 2022

Webber, Bob, email to Tracy Casselman, Seney NWR Manager, September 2006

Wildfire 101: *The Fire Triangle and the Fire Tetrahedro*n, https://www.redzone.co/2016/02/17/wildfire-101-the-fire-triangle-and-the-fire-tetrahedron/, February 17 2016, accessed May 9, 2022

Wildland Fire Management Plan, Seney National Wildlife Refuge, Department of Interior, Fish and Wildlife Service, 2008

Wildfire, Wikipedia, https://en.wikipedia.org/wiki/Wildfire#cite_note-:0-5 3, February 2022

Wildfire Today, *Federal firefighter classification without compensation–wait, what?*, https://wildfiretoday.com/2022/06/16/federal-wildland-firefighter-classification-without-compensation-wait-what/, June 16, 2022

Wood, Vivian, *A Little History of Seney, Michigan in the Upper Peninsula*, http://www.exploringthenorth.com/seneymi/history.html, 2000, accessed May 4, 2022

Yoder, Kate, Grist, *Fleeing global warming? 'Climate havens' aren't ready for you yet*, https://grist.org/migration/fleeing-global-warming-climate-havens-arent-ready-for-you-yet/, December 7, 2021, accessed April 16, 2022

Ziel, Robert, *Michigan Field Guide for CFFDRS Fire Behavior Prediction (FBP) System*, March 2015

INDEX

G

H

N

O

P

ABOUT THE AUTHOR

Gregory M. Lusk, a native Yooper who grew up in lower Michigan and southern California, has two grown sons and now lives in Hancock, Michigan with his wife Sandra. He spent the long, dry summer of 1976 helping to suppress the largest, most costly forest fire that had burned in Michigan since 1908. In early August that year, he left his normal duties as a fire specialist for the Michigan Department of Natural Resources in Marquette and was assigned as the assistant Fire Boss for the State's suppression efforts on the Seney Fire. His experiences several years earlier in Vietnam as a platoon leader were as valuable as his degree in forestry from Michigan Tech and his extensive training in forest fire behavior in the effort. Many lessons were learned by the time he left; leaves were falling, and snow was spitting by the time he got home. Long after he retired as the Upper Peninsula State fire supervisor in 1997, he dug out his boxes of information, news clippings, maps, and notes and began writing the history of the Great Seney fire. He was motivated, partly, by the thought that "those who cannot remember the past are condemned to repeat it"; with hopes that this account will help others remember and understand this small, but important piece of Michigan history.

Photo by Sandra Lusk

A PARTING THOUGHT

As a reminder to anyone in a firefighting role, safety is paramount on every fire. Over the last fifty years, some smart, experienced firefighters identified the common hazards of work on the fireline and produced a set of rules to mitigate them. ***The 10 Standard Firefighting Orders*** *and* ***The 18 Watch Out Situations*** *have been successfully used for many years and were certainly used at the Seney fire. Use of these fundamental guidelines enable firefighters to safely work in hazardous locations giving them an early warning of any serious hazard, and the ability to escape to a safety zone before a fire behavior change could cause an entrapment.*

'LCES' *stands for Lookouts-Communications-Escape routes-Safety zones. LCES, started in 1990, is a refocusing on the essential elements of the "10 and 18." The elements of LCES form a safety system used by wildland firefighters to protect themselves from entrapment by free-burning wildfires and other fireline hazards as they do their job.*

18 Situations that shout "Watch Out!"

†1. Fire not scouted & sized up
†2. In country not seen in daylight
†3. Safety Zones & Escape routes not identified
†4. Unfamiliar with weather and local factors influencing fire behavior
†5. Uninformed on strategy, tactics and hazards
†6. Instructions & assignments not clear
†7. No communication link with crew members & supervisors
†8. Constructing line without a safe anchor point
†9. Building fireline downhill with fire below
†10. Attempting frontal assault on fire
†11. Unburned fuel between you and fire
†12. Cannot see main fire, not in contact with anyone who can
†13. On a hillside where rolling material can ignite fuel below
†14. Weather is getting hotter & drier
†15. Wind increases or changes direction
†16. Getting frequent spot fires across the line
†17. Terrain and fuels make escape to safety zones difficult
†18. Taking a nap near the fireline

† = Means a firefighter fatality has occurred because of this situation

10 Fire Orders

Fight fire aggressively but provide for safety first
Initiate all action based on current and expected fire behavior
Recognize current weather conditions and obtain forecasts
Ensure instructions are given and understood
Obtain current information on fire status
Remain in communication with crew members, your supervisor and adjoining forces
Determine safety zones/ escape routes
Establish lookouts in potentially hazardous situations
Retain control at all times
Stay alert, keep calm, think clearly, act decisively

L - Lookouts
C - Communications
E - Escape Routes
S - Safety Zones

Made in United States
North Haven, CT
25 September 2026

19127603R00134